MICROMORPHOLOGICAL FEATURES
OF SOIL GEOGRAPHY

Other books by Walter L. Kubiëna

Micropedology (1938)
Entwicklungslehre des Bodens (1948)
The Soils of Europe (1953)
Die mikromorphometrische Bodenanalyse (with others) (1967)

MICROMORPHOLOGICAL FEATURES OF SOIL GEOGRAPHY

BY WALTER L. KUBIËNA

 RUTGERS UNIVERSITY PRESS
New Brunswick, New Jersey

Manufactured in the United States of America
by Quinn & Boden Company, Inc., Rahway, New Jersey

To the great expert on polar soils
JOHN C. F. TEDROW

Contents

List of Tables

List of Figures

List of Color Plates

Preface

The purpose of this book is to present a broad view of the micromorphogenesis of the soils of the earth. It is devoted to one field of soil
micromorphology, to the comparative investigation of the general nature
and development of the soil. Its conclusions are based on forty-odd years
of comparative research and scholarly travel throughout the world.

The decisive transformations of the soil as a consequence of its development are produced in its microscopic habitats. It is possible to distinguish
not only different soils of different ages, formed in different environments,
but also different horizons and subhorizons within a single soil profile.
This is because the micromorphology of each soil and each soil horizon is
very different. What we get with the naked eye in looking at a soil sample
is a kind of bird's eye view of the whole, an indistinct impression of a mul-
titude of microscopic details which make up the entity. We are immediately aware that our impression is incomplete, that what we see is merely
the outer appearance of a complex internal world. Through modern
methods of soil micromorphological research, however, specifically through
analysis of thin sections of the living soil, it is possible to obtain a very
much more detailed view of the many microscopic worlds of the soils of
the earth. I shall try to demonstrate some of the possibilities, both in the
text and in the section of color plates.

I first presented the text of this book in a series of lectures and public
conferences given between September 1966 and July 1967 at the College
of Agriculture and Environmental Science of Rutgers University, the
State University of New Jersey, in New Brunswick, and at Purdue University, in West Lafayette, Indiana. I have found it advisable to retain
more or less the form and language of the lectures but to include the
answers to questions posed by the audience during the lectures, as well as
descriptions of my personal experiences with soils and their habitats, their
plants and animals.

My lectures and what was to be the first of several visits to New
Brunswick were initiated by Dr. J. C. F. Tedrow of the Rutgers University

College of Agriculture and Environmental Science, whose special interest is polar soils, and who has therefore been greatly interested in my investigations of high mountain soils. It was not only Dr. Tedrow's great reputation as a polar soils expert that caused me to accept the kind invitation— and I may remark that over a period of some eighteen years Dr. Tedrow has worked in most of the arctic regions of North America, Greenland and northern Europe, and also in parts of Antarctica. It was also the fact that Dr. Tedrow was concerned with including micromorphological research in his investigations and in establishing a complete laboratory of thin section preparation and analysis at Rutgers. This plan was realized thanks to the help of my wife and first collaborator, Gunde Kubiëna. With suitable laboratory facilities available, it was possible to combine the lectures with laboratory exercises in thin section preparation and soil microscopy.

In view of the close collaboration of Dr. Tedrow and his coworkers, the first part of the course of lectures was largely devoted to the micromorphology of high mountain soils in different parts of the globe. Treating high mountain soils before the soils of the lowlands was of course justified by the fact that the lower soil development stages are generally more common in mountain areas.

Still another reason arose from the search for the rules governing the distribution of soil forms and soil development throughout the globe. Our world is not two-dimensional (as it may seem in most atlases) but tridimensional, and particular attention has been given to this fact. Although the great value of tridimensional presentation was fully recognized by the great Alexander von Humboldt, the method was neglected for many years. Today it has among its strong advocates the geographer Carl Troll, and it has proved enormously important in setting forth the rules that govern soil morphology and micromorphology.

Before entering upon the subject, I believe that some preliminary remarks of another kind may be in order. The first have to do with my personal attitude towards soil science and soil research. This was determined by a very early desire for travel. Dreaming, as ten-year-old boys will, I decided to become an explorer, though I could not have imagined that at only nineteen I would find myself in Central Asia, near Ulan-Ude, today the capital of the Buryat Soviet Republic. I was taken there as a prisoner, and stayed there for two years. In my free time I worked as a botanist, and I began to establish a Siberian herbarium. Four years later, I was able to bring it safely back to Europe. In my twenty-first year I was a free man in the center of Manchuria, and at the age of twenty-two I went to Hong-Kong, Singapore, Ceylon and Suez, and finally to my home port of Trieste. Back in Austria, I studied agriculture and geology at the Hochschule für Bodenkultur in Vienna; but there is hardly any doubt that my early travels and experiences determined the bent of my scientific career. I had long since become familiar with the work of Carl von Linné, the great elucidator of the world of plants, and Antoine-Laurent Jussieu, who taught the principles of a natural system of living organisms, and Alexander von Humboldt, who showed by his life and work that the most explicit way to explore nature is through travel and continuous comparative research. Soil science became my natural science, and I was to pursue it over a lifetime.

Of tremendous importance in my decision to devote my life to the exploration of soils were the later writings of V. V. Dokuchaev and the book of his great pupil K. D. Glinka on the types of soil formation, which contained excellent descriptions of soils and their habitats in parts of Russia and Siberia.

My second set of remarks has to do with the nomenclature of soils used in this book. For a variety of reasons I have continued to apply the rules set forth in my earlier books, *Entwicklungslehre des Bodens* (1948) and *The Soils of Europe* (1953). I cannot claim this system of soils as my own, for most of the names were established by earlier soil scientists, and I have introduced new names only for concepts that did not exist in the literature. What is new in my system is the application of rules of nomenclature according to the principle of priority. The rules of nomenclature themselves are not of course my invention, but rather were deduced from the mode of naming soils that had been used by the majority of soil scientists in the past. Still, I should perhaps apologize to the American reader for clinging to the above nomenclature without translating the terms into those used in the United States.

Since every main soil type (in the sense of Dokuchaev) is characterized by a distinctive fabric, I have elected to name fabric types after the soils that they typify. The soil system and the fabric system constitute a definite unit, and I am convinced that changes in nomenclature would therefore render them unrecognizable.

Furthermore, the soil system used in my comparative investigations in the field of soil genesis is necessarily a genetic system. When soils are subjected to micromorphological analysis, the genetic dependance of certain soil types upon the main types in a genetic system is very clear. This means that every main soil type has a micromorphology associated with its genesis and that other soil types that happen to be its offspring have microscopic fabrics that can be recognized as transformations of the main type. These genetic interrelations will be pointed out in discussing the transformation forms of Braunlehm. The advantages of a genetic system are very great in identifying and describing the numerous poly-genetic soils, for in these soils the climatic and environmental influences of the geological past can prove more effective than the influences of the present. In fact, new knowledge of paleopedology and the paleobiology of the soil will greatly depend on the genetic approach in soil micromorphology.

All forms in the realm of soils are subject to change in space and in time, and one of the great tasks of the soil scientist of today is to follow the changes in the two dimensions and to investigate both their causes and their rôle over the ages of soil development.

The exploration of nature by the new micromorphological methods of investigating soils has begun to open up an immense and fascinating province, where great complexity and great variety exist in microscopic space. If the era of great explorations of the unknown surfaces of the earth is closed, the age of exploration of the interior world of its soils has scarcely begun. From the Equator to the polar regions, from the great depressions to the highest mountain peaks, the immense microscopic world of soils and soil life is still almost unknown.

I have put emphasis on pedology as an independent natural science, but this does not mean that I think we can disregard the great necessity for specialized soil research to serve the practical needs of agriculture and forestry. On the contrary, I believe that the new principles and methods set forth in this book can stimulate new research and new points of view that can advantageously supplement and complete the old methods designed to serve the applied sciences. Micromorphology, based on general soil science, is already a valuable aid to comparative research. It has yielded deeper knowledge of the different soil development tendencies, and it will no doubt be indispensable in the planning of projects for the restoration of abandoned soils in the semidry tropics, the subtropics and the Mediterranean region. These practical applications are traced in the lectures as much as the general topic permits.

This book would not have been possible without financial assistance for travel and laboratory work on appropriate projects. For that invaluable help I extend warm thanks to various European institutions, particularly the Spanish Consejo Superior de Investigaciones Científicas and the Portuguese Instituto de Alta Cultura (for travels in the East Atlantic islands and Africa), to the Research Council of the Free Hanseatic City of Hamburg (for travels in South Asia, Africa and South America), to the German Research Community (for travels in Germany, Russia and South America), and to the Agricultural Research Institute of the German Potash Industry, Hanover (for travels in South America, New Zealand, the Fiji Islands and Hawaii). In addition, my thanks go to various European and North and South American universities for offering me visiting professorships that enabled me to combine lecture duties with valuable excursions and field studies in many countries.

For help in the preparation of this book, I am indebted first of all to the National Science Foundation, Washington, D.C., for awarding me a Senior Foreign Scientist Fellowship for work in New Brunswick, Lafayette, Columbus, New York City, and Point Barrow. I am also indebted to Dr. Leland G. Merrill, Jr., Dean of Agriculture and Environmental Science, and to Dr. Warren R. Battle, Chairman of the Department of Soils and Crops at Rutgers University for their kind reception and continuous help and readiness to aid in any difficulties that arose.

The most special thanks go to Dr. J. C. F. Tedrow, who not only initiated the lectures and plans for their publication but helped on innumerable occasions to make the stay of my family and me both agreeable and profitable. Last but not least, I express appreciation to my wife for her indefatigable assistance in the laboratory work and exercises. Finally, to the many other unnamed helpers who contributed to my endeavors and whose assistance was invaluable, notably in the preparation of the photographs, I extend sincerest thanks.

WALTER L. KUBIËNA

September 1, 1969
New Brunswick, New Jersey

Technical Remarks

NOMENCLATURE OF SOILS

For reasons explained in the preface, the mode of naming soils used in this book conforms to their micromorphology. I did not try to create a new nomenclature, but instead used names already established in the literature. New names have been introduced only for new soil concepts—those not yet named and described by soil scientists. The new names follow accepted rules of nomenclature, or were deduced from the generally accepted modes of naming soils.

In the figures, different patterns of pen hatching or screening serve to indicate the dominant micromorphology of soil horizons. Descriptions of soil fabric types can be found in the glossary and in the descriptions of the color plates. The color reproduction is in general true to nature. In those few instances in which it is not, the Munsell values are given in footnotes to the descriptions of the color plates.

The basic soil category, corresponding to genus in the system of plant and animal names, is the *soil type,* in the sense used by Dokuchaev (1879) (global type, Great Soil Group). The name of the *soil type* is a noun, capitalized. The *subtype* is expressed by an adjective or by another substantive, which may also be capitalized (examples: Humus-Podzol, Tangel-Rendzina, Calvero-Braunlehm, Braunlehm-Vega). The names of the varieties appear in a similar way.

NOMENCLATURE OF THE PROFILE HORIZONS

The system for designating profile horizons was chosen primarily to suit the book's purpose of treating the micromorphology of soil development in different horizons. The model diagrams of Figure 1 indicate the main horizon designations used. The designations for paleosoils, buried and redeposited soils are described in Chapter V (Figure 2).

PARALLEL PROFILE VARIATIONS IN THE SOIL DIVISIONS

Soil develops in three different realms: on land, in flood areas. and swamps, and in areas permanently covered by water. On this basis, soils

Figure 1. Soil Profiles Showing Horizon Designations

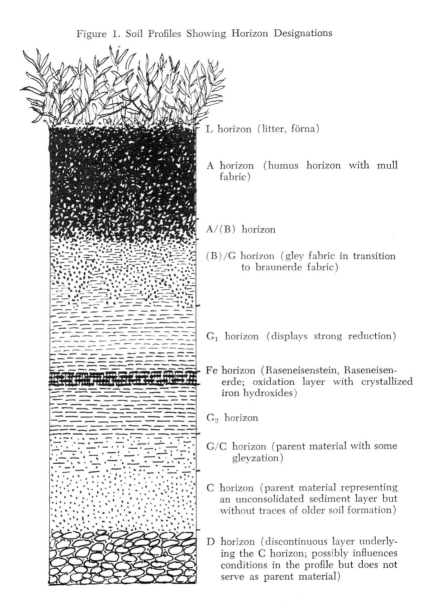

L horizon (litter, förna)

A horizon (humus horizon with mull fabric)

A/(B) horizon

(B)/G horizon (gley fabric in transition to braunerde fabric)

G_1 horizon (displays strong reduction)

Fe horizon (Raseneisenstein, Raseneisenerde; oxidation layer with crystallized iron hydroxides)

G_2 horizon

G/C horizon (parent material with some gleyzation)

C horizon (parent material representing an unconsolidated sediment layer but without traces of older soil formation)

D horizon (discontinuous layer underlying the C horizon; possibly influences conditions in the profile but does not serve as parent material)

A. Warp Soils

Note: The following may also occur: (A) horizons in Rambla, with layers that display increased soil life and root growth but do not have a humus horizon.

Figure 1 (cont'd)

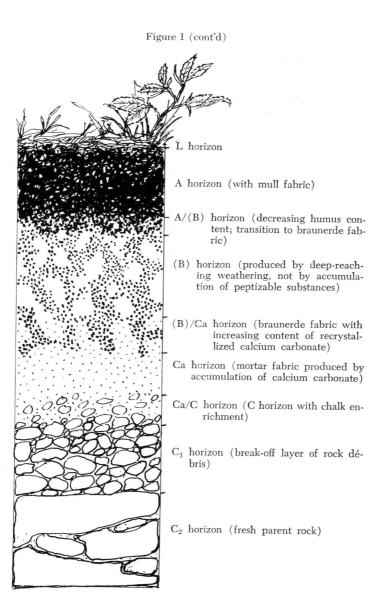

L horizon

A horizon (with mull fabric)

A/(B) horizon (decreasing humus content; transition to braunerde fabric)

(B) horizon (produced by deep-reaching weathering, not by accumulation of peptizable substances)

(B)/Ca horizon (braunerde fabric with increasing content of recrystallized calcium carbonate)

Ca horizon (mortar fabric produced by accumulation of calcium carbonate)

Ca/C horizon (C horizon with chalk enrichment)

C_1 horizon (break-off layer of rock débris)

C_2 horizon (fresh parent rock)

B. Soils of the Braunerde Region

Note: The following may also occur: (A) horizons in **Syrosem**, (B)/C horizons in chalk-deficient soils.

Figure 1 (cont'd)

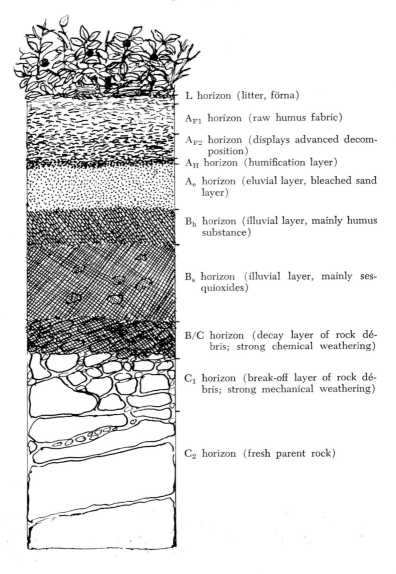

L horizon (litter, förna)

A$_{F1}$ horizon (raw humus fabric)

A$_{F2}$ horizon (displays advanced decomposition)

A$_H$ horizon (humification layer)

A$_e$ horizon (eluvial layer, bleached sand layer)

B$_h$ horizon (illuvial layer, mainly humus substance)

B$_s$ horizon (illuvial layer, mainly sesquioxides)

B/C horizon (decay layer of rock débris; strong chemical weathering)

C$_1$ horizon (break-off layer of rock débris; strong mechanical weathering)

C$_2$ horizon (fresh parent rock)

C. Soils of the Podzol Region

Note: The following may also occur: g horizons (Pseudogley layer in Molken-Podzol with local stagnation of rainwater), G horizons and Fe horizons in the subsoil of Groundwater-Podzols.

can be classified as (1) subaqueous soils, (2) semiterrestrial soils and (3) terrestrial soils.

Environmental conditions and life forms on strictly terrestric land surfaces are very different from those in areas that have a high moisture content or lie below the water table, sometimes in lake and sea bottoms. Consequently, differences in habitats and weathering, differences in humus formation and profile development, and a most characteristic soil micromorphology can be recognized in inundated areas as well as on land.

This book is devoted primarily to the micromorphology of the development of terrestrial soils, with only passing remarks about semiterrestrial soils; however, it will be apparent that the same treatment is applicable to the development of subaqueous soils and their micromorphology. Knowledge of the subaqueous soils is important to practical soil science in some countries, for many soils are of polygenetic origin and thus may begin as subaqueous soils and change to semiterrestric and finally to terrestric soils; the later phases exhibit typical properties inherited from preceding phases. (In Finland some soils have retained aqueous soil names in the terrestric stage of development—for example, the Gyttja soils of the Finnish lowlands.)

The nature of the corresponding environmental conditions and soil life becomes readily apparent as one attempts to order the main soil types of the different divisions under the principal types of profiles. To simplify the overview, only the main soil types and some subtypes are discussed in this book. Pseudogley, which is in general a transformation soil of terrestrial type, has also been omitted. The simplified scheme (a modified Marbut scheme) is intended merely to give an overview of the possibilities of profile variation within the three realms of soil formation. It is not intended as a scheme for soil systematics.

I. (A)C Soils
(Raw Soils)

Included as raw soils are types exhibiting soil life and soil processes, but without macroscopically distinguishable humus layers; these soils have only an upper horizon and only a little organic life (with or without development of plant roots).

Subaqueous: Protopedon (subaqueous raw soil)
Semiterrestrial: Rambla (semiterrestrial raw soil)

Terrestrial: alpine Råmark, polar Råmark, Yerma (raw soil of the dry desert), Syrosem (nonclimax-forming raw soil)

II. AC Soils
(*Pedosils*)

These types have a distinct humus horizon, but not Ca or fully developed (B) horizons on siliceous or silicate rock.

Subaqueous: Dy, Gyttja, Sapropel, Reed-Fen, Carex-Fen
Semiterrestrial: Paternia, Anmoor, High Moor Peat
Terrestrial: Ranker, Para-Chernozem

III. ACaC Soils
(*Pedocals*)

These types have distinct A and Ca horizons, but not a fully developed (B) horizon.

Subaqueous: Chalk-Gyttja
Semiterrestrial: Borovina, Smonitza
Terrestrial: Rendzina, Chernozem

IV. AGC Soils
(*Pedogleys*)

Such types have a G horizon below the humus horizon.

Subaqueous: Gyttja-Gley, sapropelic Gley
Semiterrestrial: anmooric Gley, Moder-Gley, Mull-Gley
Terrestrial: none

V. A(B)C Soils
(*Pedalfers*)

These soils have a pronounced B horizon, not an illuvial horizon built up by accumulation of peptizable substances, but one produced by deep chemical weathering with sufficient oxidation.

Subaqueous: none
Semiterrestrial: brown Vega, Braunlehm-Vega, Rotlehm-Vega
Terrestrial: Braunerde, Braunlehm, Rotlehm, Roterde, Terra fusca, Terra
      rossa

VI. A(B)CaC Soils
(*Pedalfercals*)

These soils have pronounced (B) and Ca horizons.

Subaqueous: none

Semiterrestrial: none

Terrestrial: Chalk-Braunerde, Chalk-Rotlehm, calcareous Terra fusca and
Terra rossa

VII. ABC Soils
(*Leached Pedalfers*)

These are soils with illuvial B horizons, but without fully developed
bleached sand layers.

Subaqueous: none

Semiterrestrial: none

Terrestrial: Semipodzol, Braunerde-Lessivé, Braunlehm-Lessivé, Rotlehm-
Lessivé

VIII. AAₑBC Soils
(*Bleached Pedalfers*)

These are soils with illuvial B horizons and fully developed bleached
sand layers.

Subaqueous: none

Semiterrestrial: none

Terrestrial: Podzol, bleached Braunlehm, bleached Rotlehm, Solod

IX. Crust and Pan-Forming Soils

Subaqueous: none

Semiterrestrial: Iron-Pan-Gley

Terrestrial: lime crust soil, Cuirass Laterite

The above scheme not only suggests the decisive influence of environmental conditions within the three realms of soil development but also
indicates the succession in profile formation by geological soil evolution.
In the earliest periods, when soils existed only in the form of subaqueous
soils, the more complicated profiles did not form. The more highly developed profiles of the soils in groups V, VI, VII, VIII and IX could form
only in the most favorable terrestric environments, the most favorable
period being the upper Mesozoic.

W. L. K.

MICROMORPHOLOGICAL FEATURES
OF SOIL GEOGRAPHY

There are few studies more fascinating, and at the same time more neglected, than those of the teeming populations that exist in the dark realms of the soil. We know too little of the threads that bind the soil organisms to each other and to their world, and to the world above.
—Rachel L. Carson

I

Principles of Soil Micromorphology: General

The methodology of micromorphological research was described by this author thirty years ago when soil micromorphology was still in its beginnings (Kubiëna, 1938). Today the technique has attained such a degree of development that a separate work would be needed to cover that subject. The methods have become manifold and vary according to needs in the different fields of soil science. This is evidenced by the great number of publications by specialists, a survey of which can be obtained in *Bibliography on Micropedology, 1955-1965* (Commonwealth Bureau of Soils, Harpenden, England, 1965); in the proceedings of the international working meetings on soil micromorphology (Altemüller and Frese, 1962; Jongerius, 1964); in *Fabric and Mineral Analysis of Soils*, by Roy Brewer (1964); and in *Die mikromorphometrische Bodenanalyse*, a recent book of the author and his collaborators (1967).

In the following chapters, only those methods will be described that the author used for comparative investigations in his most favored working field of soil development; however, it is essential to begin by describing and discussing some general principles of soil micromorphology that are common to the work of all specialists who use micromorphological methods. This will be followed by a presentation of the general rules of soil development, beginning with the principal facts of humus formation, monogenetic soil development and its changes in space (environmental pedology, soil geography) and in time (historical soil evolution, paleopedology). Finally, some facts on the micromorphology of polygenetic soil development will be discussed and will be related to aspects of practical agriculture and forestry.

MICROMORPHOLOGICAL ANALYSIS

In soil micromorphology, we proceed from the concept that the soil is not just a mixture of various constituents but is an ordered system with a construction of its own, a particular dynamics (of physical and chemical processes) and a particular biology (of soil organisms and their activity).

Because of the characteristic internal construction of the soil, the characteristic interaction of particular processes, we might compare it with the works of a watch from the point of view of someone coming from outside our world. Such a person, seeing a number of different watches for the first time, could develop a series of ways for investigating them scientifically.

(1) He could put each watch into a mortar, pound it to a fine powder and then determine the chemical composition of the whole. He would find that it is composed of a certain number of metals, each one present in a certain quantity, plus some glass, some jewels and some fine-grade oil. Of course he would not gain any knowledge of the action of driving wheels, checking wheels, springs, screws, pivots, chain links, levers, and so forth, which are essential for the operation of a watch. He would know only that watches are made of different metals and nonmetals, including glass and stones.

(2) He could take each watch apart and perform a kind of mechanical analysis by sorting the different pieces into groups. He could then determine the ratio of the sizes of these groups by a gravimetric method. Both methods would enable the analyst to classify all the watches; however, the results obtained would leave him with many unanswered questions. The quantitative data would not give him any insight into the interlocking of the parts of a watch, nor could he conclude anything about the function of each part, much less the functioning of the whole.

(3) He could leave each watch intact and investigate each part in its place, determine its position, its connection with other parts, until he had a complete knowledge of the construction of the whole.

(4) He could investigate each watch in a state of motion and observe directly the movement of each part and its individual rôle in the function of the whole.

In terms of soil research, only modes (1) and (2) were applied in the past. A single exception was in the investigation of profile morphology. The principles of soil micromorphology as a special working field are similar to those in modes (3) and (4) exclusively. The principle of "undisturbedness" is observed, and the constituents are investigated in their relation to each other and in regard to their rôle in the function of the whole.

The soil is more than just a kind of clockwork, however. It represents a whole world similar to our own, although it is a world we do not live in ourselves. It is an organized space of life like the space in which human beings live, a "living system" as S. A. Waksman put it, a microcosm.

For the micromorphologist, the only reality of his working object is the undisturbed and living soil in the natural environment. All other approaches to soil science, however important and necessary they may be, represent makeshifts. Only living soil allows direct observation: the investigation of its natural changes. In spite of this, soil micromorphology cannot limit itself to the investigation of soils in their natural environments. Many details can be recognized only by application of micromorphological laboratory methods. For this purpose we need to take soil samples with a soil knife in special sampling frames (Kubiëna, 1953a).

Nonetheless, a soil sample really is not soil. It represents the soil as leaves and flowers in a herbarium would represent the tree itself in its natural environment. A soil sample when dried undergoes no further changes, has no dynamics, no life. Even so, many details are preserved in it if the natural structure has been left undisturbed. Every soil, every horizon of it, has a typical micromorphology (soil fabric) in which the result of its dynamics, biology and development until the time of their interruption is clearly pictured. There are only a few ways to recognize a soil as an entity, using simple methods. They are the following: (1) determination of profile morphology; (2) determination of the humus form; (3) investigation of the micromorphology.

The soil profile, the humus form and the micromorphology are the most typifying characters for every soil. Soil scientists generally investigate the first two, but as yet not enough is known about the third. For the progress of soil science in general, knowledge of the soil as a whole and of its position in the general system of soils is of enormous importance. In my experience, micromorphological analysis, used in conjunction with methods for determining profile morphology and humus form, provides the most complete insight into the essence of a soil. And there is no simpler or faster method than micromorphological analysis to achieve this end.

THE PRINCIPLE OF COMPARATIVE RESEARCH

This is another research principle that developed from micromorphological analysis. Every phenomenon in the microscopic realm of the soil depends upon the life and environmental conditions of a habitat. The various phenomena in a soil may occur in different places in the world with different degrees of intensity and in different stages of development, but they always occur in a way characteristic of the habitat and its environmental conditions. The fact of the characteristic variability and its significance for soil diagnosis, and the great diversity of soil micromorphology, bring the necessity for travel, for direct observation and comparative analysis of the different soils that exist in various environments, for the search after the similar and the contrasting, and for the determination of the decisive morphological characters of the soil types. We have to follow the same line that Linné laid out in the early days of botany and zoology in exploring and classifying the great morphological multiplicity of plants and animals. To this purpose, we need not only to investigate the soils in their native environments but also to establish collections of framed samples containing all horizons of the investigated profiles for future research. For the same reason, we need a collection of thin sections and a collection of color slides of the different soil fabrics, soil profiles and soil habitats.

Many soils are formed primarily by present environmental conditions, but others may be products chiefly of the past. Micromorphological analysis is the only method that allows us to differentiate clearly between the characters of former and present influences in the interior of the soil, for it is governed not only by geographical principles but also by historical considerations.

II

Principles of Soil Micromorphology: Differential Diagnosis

It has been stressed that micromorphology treats the soils as entities, as undisturbed natural bodies. Every soil phenomenon in its microscopic realm is formed by the life and the environmental conditions of its habitat; this is a biological habitat because the soil layer, the pedosphere, is a part of the biosphere. The soil micromorphology is extremely varied, but is always typical of the habitat. This is why one of the main principles of micromorphological research is comparative analysis, the search for the similar and the contrasting—and the finding of the typical in the sense of the biological sciences.

In the course of micromorphological research, the soil begins to appear to be something not taken into account previously—a natural body of great heterogeneity. The soil scientists of the past knew this but had no method for pursuing the subject. I quote from the literature: "The soil is an extremely complicated system, a factor which in its essence will always remain unknown." "The soil is full of unfathomable mysteries." "One could dedicate a whole lifetime to the investigation of a single soil and at the end still not really know it."

How can highly heterogeneous natural formations be distinguished from one another? There are two ways:

(1) By looking for measurable properties that all varieties have in common but possess in different quantities (from zero to x). Examples of such properties in soils are calcium carbonate content; acidity; organic substance; sand and clay content (texture).

All properties of this kind can be designated as homogeneous quantitative properties. The exactness required to distinguish different soils from one another can be seen in the exact measurement of the quantity of a property or constituent. The disadvantage of this method is that the number of measurable characteristics, which may represent homogeneous quantitative properties, is not very great. In a general distinction of soils, usually not more than five or six are in use. These are sand and clay content (texture), organic substance, pH, calcium carbonate content, pore space content, and water-holding capacity.

Some of these properties are not really homogeneous; the property "organic substance," for example, is subject to the greatest diversity and differences in value.

(2) By looking for those characteristics by which the various natural formations differ most from one another. With this method the problem no longer is quantity (the amount of a certain constituent); greatest diversity will be demonstrated by the simple fact of the "invariably present" on the one hand and the "invariably absent" on the other. These properties or characteristics are designated therefore as heterogeneous qualitative properties. The disadvantage of this method is that the investigator has to deal not with just a few but with an uncountable number of properties within a single formation.

How to approach such an enormous number of properties? The method by which this is done may be designated as differential diagnosis.

By this method, soil formations are compared with one another as entities, and are investigated for those details or properties by which they differ most from one another. For each one of them, these properties are the most characterizing, the most typical. The result is that we do not seek five or six properties common to all soils, although present in different degree, but instead some highly specialized properties that are completely different in each soil.

GENERAL APPLICATION OF THE TWO PRECEDING METHODS

The first method, differentiation and diagnosis by a small number of homogeneous quantitative properties, was developed and applied successfully with more or less homogeneous industrial products, such as steel, glass, superphosphate, sodium nitrate, and so forth. This is a kind of product-testing. This method was used exclusively over long periods of time in soil research. It was developed under the influence of classical agricultural chemistry in the early years of soil investigation. It has proved fruitful and is still used today and will continue to be used in the future. The only mistake has been to rely on this method alone, neglecting any others that might reveal the great multiplicity of the soil.

The second method will always be indispensable for investigating formations with high heterogeneity, such as plants, animals and soils. Differentiating between various plants and animals only by the data of their chemical composition could not prove entirely satisfactory, although these and other quantitative measurements yield important information. M. Rubmer in his *Handbook on Human Anatomy* cites some data on the differences in composition between the human male and the human female:

	Male %	*Female* %
Muscle substance	41.8	35.8
Fat	18.2	28.2
Skeleton	15.9	15.1
Other substances	24.1	20.9

These findings in the field of quantitative anatomy could profitably be completed by chemical analyses of the content of fats, proteins, chondrogens, collagens, calcium carbonate and calcium phosphate, or of the production of the different hormones. But would it not be disastrous if we had no other methods for distinguishing between males and females?

Soil chemistry and soil micromorphology really are not far apart, however. In essence, soil micromorphology is a kind of soil chemistry, just as thin section petrography is a kind of detailed rock chemistry. The soil micromorphologist must know the chemical composition of the fabric elements he is dealing with if he wants to understand their rôle in the function of the soil, in its dynamics and biology, and the causes of their particular morphology. Soil micromorphology as such represents a valuable auxiliary working field for the soil chemist since it allows him to control his results (by another technique).

Substances occur in the soil in very different forms and with different properties, but only in forms and with properties that are characteristic of those given constituents. Thus, from the forms and microscopic properties, the investigator can draw conclusions about the composition of the fabric elements.

INVESTIGATION OF THE FIELD-EFFECT (FELDWIRKUNG)

The soil is not just a mixture of constituents. Nor is it the sum of its constituents or the sum of its properties. What, then, is it?

(1) The constituents must be in combination with each other in a characteristic way to make a soil. Therefore, a soil sample taken without a soil frame is nothing but a heap of ruins, a pile of débris.

(2) Each constituent is not only a substance but, as a constituent, plays its rôle in the dynamics or biology of the soil as an entity.

(3) No constituent plays its rôle independently; all influence the others, to a greater or lesser degree. Relations exist among the separate constituents, and between the constituents and the whole.

(4) The composition of the microscopic plants and animals which are a part of the soil, and which play the main rôle in its function, is not something accidental. This, too, is typical for each soil. The organisms of a soil are not just a mixture; they have developed into a kind of assemblage, which we might speak of as an assemblage of species (*Artengefüge*).

(5) If we regard biological processes as processes and not as manifestations of life, then we have a kind of assemblage or a fabric of processes typical of the biology of a soil. A similar assemblage can also be formulated according to the physical and chemical processes, that is, its dynamics.

(6) No factor in the entity of a soil acts independently; all are influenced by other factors. The reality of the soil is not the sum of the action of the single factors; it is the interaction of all factors in unity.

As a result of all these statements, we are led to realize that the so-called field-effect (*Feldwirkung*), one of the main principles of Albert Einstein's conception of the world, must be valid also for the soil, and must be of considerable value for soil research in general.

By a field (*Feld*) (after Einstein) is understood a totality of facts, present at the same time, which can only be understood by their mutual dependence. After the relativity theory and the theory of quantum mechanics, the concept of the field-effect is regarded as the third of the basic pillars of our present world conception. The concept of the field-effect will, in my belief, substitute for most of our present concepts and approaches to practical soil science research, since these are now based on the study of isolated single properties. The time will come when the old methods will give way to new ones devoted to the exploration of the field of factors which are responsible, as a whole, for the phenomena produced by them. This period has already begun, although many of its initiators may not yet have become conscious of it.

In my study of the influence of the soil on the development of the Panama disease of bananas, the field-effect became clearly evident (Kubiëna, 1967). Earlier investigators repeatedly believed they had isolated the factor responsible; for example, the grain size ratio (texture), the nutrient content of the soil, its pH, the presence of free calcium ions, and so forth. But there were too many exceptions to the assumed favorable or unfavorable effect of these single factors to allow credence to their findings.

In the early 1960's my colleagues and I had an opportunity to investigate a great variety of habitats in tropical banana plantations (Gros Michel variety) of Liberia, Fernando Póo, Colombia, Ecuador and Honduras and we found a typical variation of the field of responsible factors. The fields and their effects changed with the different soils.

The marked influence of the soil as a whole on the establishment of different fields of effective factors seems to apply to soil fertility. At one time the nutrient content of the soil was used exclusively as the basis for determining the amount of fertilizer to be applied, but it was found that the same amount of fertilizer applied to different soils had different effects even where the soils had the same nutrient content. This resulted in various papers which described the influence of different single soil factors on the effectiveness of a given fertilizer. Factors discussed were the influence of moisture conditions, texture, pH, calcium content, air content, humus content, humus form, and so forth. However, it is the influence not of a single factor (although some factor will always be paramount) but of a particular field of interdependent factors that makes the actual difference.

In reality the whole soil in its natural environment is the decisive factor—the unity of all possible influences. This is why the subjects of soil fertility, soil classification and soil mapping have a close affinity. All methods that serve to ascertain the formation of the soil as an entity—including micromorphological analysis—will become indispensable to the specialist in soil fertility. Many of the decisive factors of the field can be recognized directly from micromorphological investigations of the undisturbed soil.

III

Influence of Life and Environment on Soil Micromorphology

It is said that life has a very particular influence on soil formation. What kind of influence is this? The French writer Anatole France wrote in one of his books: "La vie n'est qu'une moisissure superficielle qui n'altère pas la pureté minérale des astres."[*]

In essence this is sarcasm and refers primarily to human life, but if we take it literally, there remains a well-formulated statement on the effect of life in general, as much as we know about it from our limited horizon. We know that our biosphere is very thin. We know that on the largest school globe it could not be represented even by a film of breath, so thin it is. And we know that this biosphere is the only space of life with which we are acquainted, and that the rest of the world, inaccessible to us until now, is unanimated, uninhabitable and mortal.

Still, the statement of Anatole France permits the conclusion that the influence of life upon the mineral world must be considerable. Think of a kind of life which exists only in the form of a superficial mold, but a life which, if it were to penetrate deeply into the mineral purity of the stars, would completely alter the stars.

We are afraid if a mold attacks our food, if it invades a glass of marmalade or a piece of bread. In the tropics we are afraid of molds on our shoes and books. It happened not long ago that a trunk of books and manuscripts that I had shipped to New Brunswick fell into the sea during the loading. I received the trunk, finally, full of green and yellow molds which had transformed my literature and manuscripts into a pile of manure. We see, therefore, that it is not the mold as such we are afraid of; it is the series of changes it produces in the substances attacked.

In much the same way mold grows on a piece of bread, life grows on the surface of the solid earth crust. What kinds of changes are produced by its development? When and in what way does it become a soil? We

[*] "Life is nothing but a superficial mold which does not alter the mineral purity of the stars."

have a method that allows us to control this kind of inquiry in every detail: this is the thin section method. We can prepare any material in a thin section, from hard rock to completely loose organic substances, weathering products, organisms and organic residues. Each remains in its original place, undisturbed, and the whole is ground down to a thin plate of three hundredths of a millimeter (Plate 1).

The development of a soil begins with a bare rock and continues through a sequence of different genetic stages until a definite final stage is reached. This we call the climax soil. The soil development stages, in general, parallel the development stages of life. The pioneer stages are reached very rapidly, so that bare rock surfaces scarcely exist in nature. Everywhere, colonization may be produced by threads of algae, colonies of diatoms, hyphae of fungi, colonies of bacteria with protozoa, nematodes, and other organisms; together they constitute a kind of soil life, although a primitive one. By micromorphological analysis we can recognize the first phenomena that are characteristic of soil development: accumulation of different forms of decomposition products, of organism remnants, formation of humus substances, beginnings of biological weathering of the mineral substances.

The next stage is the colonization of the rock surfaces prepared for by crust-forming lichens which continue making the substrates suitable for further colonization (Plate 2A).

The crustose lichens (particularly under forest cover) are followed by leaf-producing frondiferous lichens and bush-shaped arbuscular lichens. Between the little leaves and brushes they accumulate not only detritus but a great many mineral particles transported by the wind. In these layers a marked development of the mesofauna is produced, largely by the activity of small arthropods such as horn mites (oribatids) and collembola.

These lichens soon become overgrown by different moss varieties in which, under sufficiently humid climates, develop different humus forms. These, in their fabrics, already correspond in every respect to those of well-developed soils which have an organized soil life (Plate 2B).

The accumulation of rock detritus and detritus of organism residues within the cushions of the moss cover is considerable. Under the growth of higher plants, which gradually follows, soil development becomes much accelerated, particularly where there are Gramineae. The grass cover not only collects considerable quantities of mineral and organic detritus but fixes it and produces higher humus forms such as mull-like moder or moder to mull (Plate 4E and 4F). The soil forms aggregates and, finally, coherent sponge structures. Plant residues from grassland covers represent a very favorable raw material for the formation of valuable humus forms. Natural grassland cover spreads and gradually covers all bare rock surfaces.

Below the grassland cover, physical and chemical weathering of the underlying rock is greatly intensified. This intensification is caused particularly by the widespread system of rootlets which penetrate even the finest cracks and intergranular spaces of the parent rock. Thus, the beginning of the formation of a B horizon becomes visible in time. This is a slow process. Whereas the formation of humus horizons sometimes can take place within a few years, the formation of B horizons requires decades to several hundreds of years, depending upon environmental conditions.

SOIL LIFE AND THE MINERAL PART OF THE SOIL

From micromorphological analysis, it is clear that soil life is responsible not only for the decomposition of organic substances and their transformation into humic substances but also for the production of a number of other important phenomena that are very characteristic of soils as natural formations. These phenomena are:

Mixing of the organic constituents with the mineral constituents, which is accomplished mainly by the action of small arthropods.

Formation of loose moder aggregates (in mull-like moder) of mineral and organic constituents even if very little clay or no clay at all is present. This formation is produced mainly by larger arthropods, such as julids and glomerids (both diplopods), wood lice, different insect larvae and some earthworm species (Plate 2F). The humus forms produced are mull-like moder and mull-like rendzina moder.

THE CLAY-HUMUS COMPLEXES

By the action of certain soil animals, in the first place by some earthworm species, some kinds of complexes of humic substances and inorganic weathering products, mainly clay minerals, amorphous iron hydroxide and hydrated colloidal silica, are produced. Their constituents cannot be separated by micromechanical methods, and only to a certain degree by chemical methods. The nature of these complexes is not fully explained. Micromorphologically, they can be clearly distinguished from other formations. Their important rôle in the formation of water-stable aggregates and sponge structures can also be clearly recognized (Plate 4F).

LIFE AND AGGREGATE FORMATION

The aggregates and sponge structures are products of animal activity and are caused mainly by species of earthworms and enchytrae. Thus, life plays an important rôle in the genesis of soil fabrics and aggregate formation. Soils are formed to a large extent by the environmental conditions of a habitat; but without the influence of life, well-developed soils are inconceivable. Therefore, soils promote and produce life, but they are also products of life.

Every living creature—not only the organisms that live in the soil but also those that live upon it and from it—plays its rôle in the formation of soils. This includes man. No organism living on earth escapes its destiny of becoming parent material in the formation of soils. When we ourselves die, the soil that has nourished us over a lifetime claims us back into its natural rotation, regardless, as we might say with a little sarcasm, of whether we have cared or not about the facts and rules of its existence and formation.

INFLUENCE OF ENVIRONMENTAL CONDITIONS OF THE
HABITAT ON THE SOIL MICROMORPHOLOGY

A most effective factor in the soil-forming process is the air climate (rainfall, temperature, and radiation of the sun, with seasonal alternation).

The soil climate, although influenced by the air climate, may differ from it considerably. The soil climate represents a very specialized soil characteristic, and is subject to typical seasonal alternation in humidity and temperature. Of particular influence is the presence of winter frost, permafrost or frost-free conditions under snow cover (the last of which allows soil life to continue its processes in winter). The micromorphology is greatly influenced by the soil climate, or even the climate of a soil horizon. But the soil climate may be greatly influenced also by the micromorphology. Thus a soil horizon may become the creator of its own climate.

The effectiveness of a sponge fabric in this respect has been brilliantly demonstrated by direct humidity and temperature measurements in the interior of the various horizons of a forest soil by one of my former collaborators, G. Zachariae (1967). Dr. Zachariae is a soil zoologist who uses micromorphological analyses to determine the efficiency of the activities of different groups of soil organisms. He combines these analyses with meteorological measurements in the natural habitat. For this purpose he uses special thermocouples particularly adapted to his investigations and, in addition, time-recording instruments to measure hourly temperatures.

The soil layer investigated by Zachariae consisted of a humus horizon rich in cavities which displayed a much more intensified decomposition and humification than the soils of the other forest habitats of the region (the German Black Forest). Favorable conditions of soil climate create an excellent soil life characterized by a highly active bacterial flora which makes the plant residues (mainly spruce needles) easily assimilable for the well-developed earthworm fauna. The earthworms create the sponge fabric with its many spaces. Investigations with specialized apparatus showed that the humidity in that soil is not caused solely by the percolating rainwater, but that it is considerably increased by the condensation of water from the saturated soil air on the soil pore walls during the night and morning hours. This condensation in the soil results from even a slight drop in the temperature aboveground, even though the latter may be well above the dew point.

Moisture conditions independent of the air climate also influence the micromorphology of the soil, for example, water influx, water flowing away, water conductivity, water-holding capacity, and stagnation. The surface relief and the profile morphology are particularly influenced by the presence of different kinds of subsoil such as sediments of clay, limestone, gravel and sand.

Soil life, soil climate and water conditions influence the kind of micromorphology to such a degree that by comparative thin section analyses of soil fabrics and fabric details it is possible to ascertain the environmental conditions of a given soil or even of different soil horizons.

The micromorphology of the soil is greatly influenced by the iron, calcium carbonate and general nutrient content of the parent rocks, including its grain sizes, pore space content, weathering resistance and binding substances.

The soil also acts as its own environmental factor, since many of the soil-forming conditions depend upon its composition and degree of development. A good example is Chernozem. Its thick humus layer and the preservation of its calcium carbonate prevent a deep weathering and the formation of B horizons. The so-called Atlantic Ranker on granite in northern Spain displays a similar influence. Its great water-holding capacity and the great water consumption by the dense and active soil life impede a deep chemical weathering.

THE INFLUENCE OF MAN

Man is a living organism in nature, and his activity on the soil resembles the activity of other living organisms which, like man, can change completely the character of a soil. Organisms which exert that kind of influence include the gophers, the ground squirrels and the moles, and, with even more effectivity, the earthworms, the ants (Kalahari Desert) and the termites.

We may therefore classify the influence of man as a kind of life influence. But since we are human beings ourselves, and since we build up our science for our own use, we separate our activity from that of our fellow organisms and study our influence separately. Moreover, we can separate our modes of influence upon the soil into biological methods (manuring, mulching) and nonbiological methods (plowing and harrowing).

The development of special quantitative methods for determining the changes in pore space content, in aggregate formation, in ratio of aggregate types, in pore space types, in aggregate and pore sizes, and in the interior surface of the soil was made possible by micromorphometric analysis, which is mainly a photometric method. We recently finished a book on this subject (Kubiëna, et al., 1967).

The influence of man has not been, in all cases, favorable to the soil. There are many soils, particularly in the subtropics and tropics of alternating wet and dry seasons, and in the Mediterranean countries, which have become completely unproductive and now constitute vast areas of wasteland. Having given particular attention to the changes of micromorphology in these soils, and to the micromorphology of their improvement, it is my conviction that there is great possibility of restoring these soils. India, Brazil and vast parts of tropical Africa, where these soils occur, are precisely those regions where hunger, malnutrition and unbalanced nutrition are most prevalent. Hence the problem of restoration of these soils has become one of the most urgent ones for practical soil science (Primavesi, 1966).

IV

Soil on Limestone in the Vienna Woods:
The Rendzina Stages

Investigations in this region were performed with the Viennese zoologist W. Kühnelt (1950) and his collaborators. The development series on limestone and dolomite in the southern Vienna Woods was selected for presentation because it is very suitable as a standardized series for comparative research. Also, here, colonization of bare rocks may begin with the development of algae, diatoms, fungal hyphae, bacteria, protozoa, rotatoria, etc., in wet or temporarily wet areas of the rock surface, or in cracks and fissures. Flora and microfauna found here bear some similarity to the life in the adhesion water films and the capillary water in the cavities of soils.

Of decisive importance, however, is the colonization that is followed by crustose lichens. On limestone in many parts of the world in temperate, cold and mediterranean climates, the first lichens are chiefly endolithic lichens. A lichen is generally thought to reflect a symbiotic relationship between an alga and a fungus. An endolithic lichen penetrates some millimeters into the rock, and is covered by a crust of calcium carbonate. Thus the lichen itself is scarcely visible on the surface of the rock. However, if one scratches the rock surface with the edge of a hammer or with the point of a soil knife, a bright-green scratch is produced. The scratching renders the layer of algae visible. Very small snails of *Chondrina avenacea*, living on the rock surface, feed on lichen. With their radulae, the hard parts of their mouths, the snails are able to scrape the lime crust away and to gain access to the layer of alga. The endolithic lichen that occurs in the southern Vienna Woods is *Verrucaria calciseda* (Plate 2A). In the surface crust we observe small holes which are the openings of fruiting bodies, the apothecia, of the fungus. These are the only signs visible from outside that the lichen is present.

The *Verrucaria* has a considerable influence on the preparation of the surface layer of the rock for further colonization. When calcium carbonate is partly dissolved, leaving small spaces in the rock, residues of the lichen are deposited in them.

Further colonization takes place first of all with the development of a different form of lichen which spreads over the areas prepared by the *Verrucaria*, but which grows visibly on the surface of the rock (epilithic lichen). The epilithic lichens are primarily *Lecanora* varieties, which have a striking appearance owing to their black color. In the case of our studies, it is mainly the species *Lecanora crassa*. The rock is now softened, loosened and interspersed by more organic residues, which give a good substrate for the growth of mosses.

Almost everywhere in Europe *Grimmia* species occur. In the Vienna Woods, the stages of development investigated are found in sunny forest clearings which dry out considerably in summer. Because of this, the moss cushion, which is generally gray and looks like mouse fur, turns to pale green only after long rainy periods. The species is *Grimmia orbicularis*. Because of the softening of the rock layer underneath the lichen crust of the *Lecanora*, the root-like rhizoids of the moss can penetrate the adjoining sections on the brim of the moss cushion. Thus, the lichen is attacked by the rhizoids before the margin of the cushion has advanced towards the *Lecanora* area. Our thin sections show every detail of this development. Finally, all space formerly occupied by the *Lecanora* will be completely replaced by rhizoids of the *Grimmia* (Plate 2A).

Inside the *Grimmia* cushion, which may spread out over a considerable area, we find a typical fauna developing—small arthropods whose droppings fill out the spaces between the stems and leaves of the moss. They are cylindrical in shape and almost black. In addition to these droppings, some rock detritus is accumulated. Thus we can describe the following constituents: (1) plant residues bitten through (by animal activity); (2) mineral grains (fragments of the limestone rock); and (3) small arthropod droppings (about 60μ long, and 40μ in diameter). The whole mixture is completely loose and without aggregate formation.

In the course of the last chapter we spoke of a humus form of that fabric, calling it a moder. Because this moder developed on limestone, it contains a considerable amount of limestone fragments and is also characterized by a certain calcium content in its humus substance which produces in part a humus formation of a blackish color; we call this humus form a rendzina moder or a chalk moder (Plate 2B). In habitats with even drier environmental conditions, like the limestone area of central Spain, a formation of rendzina moder inside the *Grimmia* cushions (the species here is *Grimmia pulvinata*) cannot develop. A similar but slightly more primitive type of life is present, however. The humus form associated with it is designated as raw soil humus (not raw humus) or, to use a Russian expression, syrosem humus.

We can classify different environments by the effect environmental conditions have on different developmental stages within the initial phases. The difference in the humus just described is produced by the greater dryness of the climate of central Spain. Some other differences are of course caused by greater humidity, or by another range of temperature produced by winter frost or even frost nights in summer, as happens in high mountain regions or in the Arctic. Using the development sequence of the Vienna Woods as a standard, we can readily ascertain differences in habitats by comparative research.

ANIMALS ACTIVE IN THE MODER FORMATION OF THE GRIMMIA CUSHIONS

The method applied by W. Kühnelt in determining the principal animals involved in soil development, and responsible for typical changes in micromorphology, was an investigation into the soils in their living state and a careful isolation of the most important animals found, by direct microscopy. They were kept alive in containers supplied with some soil and their natural feeding materials.

The animals principally responsible for humus formation were oribatids (horn mites). It was significant that collembola, very frequently found in moss cushions in forests, were not present or occurred only during long rainy periods. An explanation of this is the fact that the *Grimmia* cushions in the forest clearings of the southern Vienna Woods dry out completely in dry weather periods. This would kill the collembola, which need humidity to live, and die out completely in dry air. The movements of the oribatids observed in the natural soil were very slow; it took seconds for them to raise one leg, put it down again and decide to move another. This slowness of movement is attributable to their great need to conserve water for their life processes.

The most effective species of oribatids were the following: *Camisia biverrucata*, *Scutovertex minutus* and *Trichoribates trimaculatus*. Droppings in rendzina moder are especially well preserved in their cylindrical forms. Rendzinas are therefore particularly suited for dropping analyses—that is, to find out from the size and the form of the droppings what animals are involved in the transformation processes. Micromorphology employs a method of pursuit similar to that of the forest ranger, who looks at droppings to draw conclusions about the life habits of his game animals. In German, the word *Losung* is the term for excreta in both humus micromorphology and game ranging and hunting.

We have seen that a definite humus form is found in the cushions of the *Grimmia,* just as in a soil. Is it a soil? The agriculturist would not call it a soil. The higher plants of the Vienna Woods evidently are of a different opinion. They choose it as a substrate for their growth, particularly if the moss is spreading and its rhizoids are penetrating the cracks of the limestone rock. It has thus become a substrate capable of supporting the growth of higher plants—a decisive characteristic for a soil in many definitions.

Among the higher plants that can be found rooting in the cushions, one genus is of special importance for soil development. It is the genus *Globularia.* In different environments we find different species. In the clearings of the southern Vienna Woods it is *Globularia cordifolia.*

As a consequence of the development of the *Globularia,* the *Grimmia* is overgrown and dies. In the struggle for life, other plants can also destroy the *Grimmia*—lichens like *Caloplaca fulgens.* The lichen strangling the *Grimmia,* shutting it out from the sunlight, can be seen clearly in our thin sections. The *Globularia* grows in the form of a horizontal trellis along the surface of the rock. Since the heart-shaped leaves are oriented parallel to the rock surface, the plant forms a kind of cover. Under the twigs and leaves, great amounts of loose rock débris and large rock fragments and sand are accumulated and protected against further erosion.

Among the deposits of loose rock débris is sufficient space for the life of soil animals, which fill it out with their droppings. There is no doubt that we have to consider this kind of soil a Rendzina; since it is a primitive Rendzina, we call it a Proto-Rendzina.

A similar development can be observed with higher plants, first with some grass species. The first grasses that appear in the moss cushions are *Stipa** and *Sesleria* species. In Austria, particularly dense grass cover is produced by *Sesleria varia* in this environment and is here designated blue grass. Blue grass is a most interesting plant for development studies.

* In the southern Vienna Woods primarily *Stipa pulcherina.*

Under the tufts of *Sesleria varia,* the humus formation begins in a way similar to a rendzina moder formation, but with less humification in parts with a low content of calcite fragments.

The calcite content is very important for the rendzina moder. We distinguish different varieties of Proto-Rendzinas. In humid climates the calcite fragments in the humus horizons of the Rendzinas are present only as rock detritus. Calcite grains of this kind are called clastic calcite (Plate 2C).

Soils with rendzina humus are also found in dry climates—except in those areas where conditions produce soils without any formation of a humus horizon (calcareous raw soils, Syrosems, "white Rendzinas"). The AC soils in dry (summer-dry) areas are called Xero-Rendzinas. Their biology is different, and, as a result, they are different in the intensity of decomposition and humification. They are also different in another respect—in the form of their calcite. They contain, in addition to clastic calcite, recrystallized calcite, calcite which is precipitated out of the soil solution.

The presence of predominantly recrystallized calcite in the upper humus horizon is a typical characteristic of the Rendzinas in dry climates. The reason for this is that dissolved calcium carbonate, in a humid climate, is easily washed out in the form of calcium bicarbonate as the soil solution moves downward into the soil profile. In a dry climate, the calcium carbonate is only partly leached out, and it moves distinctly upward with the rising soil solution. The particular rôle of calcium carbonate in the fabric of alpine Rendzinas will be treated in a later chapter.

MULL-LIKE MODER BELOW SESLERIA VARIA

Under the grass cover of *Sesleria,* the soil development continues to improve. Out of the rendzina moder there gradually develops a humus form that has marked aggregate formation. The humus layer deepens, the living conditions improve and the soil life is richer in species and in individuals. Chemical weathering is still slow, and inorganic material consists of loose calcite grains most of which have a grain size of coarse to fine sand. The aggregates are droppings or fragments of droppings of larger arthropods. The binding substances in the aggregates are of organic nature, mainly humic substances with a certain calcium content.

Previously, a humus form with low clay content and with aggregate formation was designated as mull-like moder (because of the mull-like appearance it presents to the naked eye). Such a humus form, rich in calcium carbonate (developed mostly on limestone, dolomite or gypsum), is designated as mull-like rendzina moder—the soil type as mull-like Rendzina.

The essential characteristic of the aggregates is that they are rich in minerals, that is, in grains of calcite or dolomite. The arthropods which produce these aggregates are able to take up with their food a considerable amount of limestone sand and bake it into their droppings. The arthropods involved in this formation belong to the glomerids* and julids.† To some extent insect larvae may also be active; in the southern Vienna Woods, the larvae of the rose beetle *Cetonia aurata*, for example. All produce droppings that are typical in shape and size, so that when the animals which took part in the transformation were carefully isolated and their excreta observed, they were easily identified.

The principal glomerids were *Glomeris hexasticha*, a drought-resistant species, as the main form. Their droppings are about 1.8 mm. long, a blackish groundmass interspersed with white calcite grains. The form of the droppings is conical, with a kind of helve at the broader end. The principal julids were *Cylindroiulus friesius* and *Chromatoiulus uniliniatus*, both remarkably drought-resistant species. Their droppings are about 1.5 mm. long and have an inlet at the middle.

Of particular interest was the activity of the frequently found rose beetle larva. Its droppings, cylindrical with a smooth, rounded surface, reach a length of about 5 mm. They contain fine and coarse rock fragments, fragments of little-decomposed plant residues rich in lignin and cellulose (Plate 2F).

DEVELOPMENT OF MULL-RENDZINA

Gradually, mull-forming earthworms multiply and become very active as more and more clay is freed by solution and accumulation out of the limestone. In habitats where clay can accumulate easily because of favorable topography or where the rock contains a large amount of clay among its insoluble constituents, this development is faster and easier. It is particularly accelerated with marl rocks.

---

* The glomerids belong to the Myriapoda, a group of the Diplopoda, which possess segments, each having two pairs of legs. The glomerids are able to roll up into a ball like some wood lice.

† Julids are cylindrically built, elongated diplopods.

The change in composition and intensity of life is effected by the gradual development of true mull as a humus form.

Humus of the Mull-Rendzina differs from the preceding kind in that the decomposition of organic residues is much intensified. The plant cover can be either a dense grassland or beech forest vegetation. The humus layer has considerably thickened, from about 12 to 23 cm. This and the increased clay content and the formation of mineral-humus complexes, together with the formation of a water-stable sponge fabric, cause a considerable improvement of the water conditions. Not only is there higher moisture capacity, but also there is higher moisture retention.

All these conditions bring about a more favorable soil life, including bacterial activity, which depends on humidity and safety against dryness. The animal population seen in former soils disappears or is diminished. The most important animals are the mull-forming earthworms, which tend to deepen the humus horizon into the subsoil; the latter is a soft loamy layer rich in calcium carbonate occurring in recrystallized form.

Because of the humid climate the leaching is intensified. The accumulation of lime gives rise to a white horizon, the Ca horizon. The dark-gray color of the A horizon passes gradually into the whitish subsoil. Under the soil microscope we see that the dark color in the transition horizon is limited to the fillings of the earthworm channels. These fillings are earthworm casts. They are deposited like the earthworm casts which fill the krotovinas of Chernozem soil. The earthworm casts are stained by humus substances, but show no plant remnants. Not only is decomposition rapid and intense but humification, the transformation into humic substances, and a kind of binding to the mineral weathering products, the clay substance, the free iron hydroxides and the free amorphous silica are accomplished at the same time.

DEVELOPMENT OF TANGEL-RENDZINA

In the course of further development, the series of Rendzinas is left behind, and different development stages with ABC profiles take their place.

A development of a distinctive kind takes place under special environment conditions in parts of the southern Vienna Woods. On the crests of the highest hills, on limestone or dolomite, at altitudes of 730 to 900 m. (2,400 to 3,000 ft.), there develops a dense plant cover of the bright-red flowering heather *Erica carnea* under forests of *Pinus nigra austriaca*, the

Austrian black pine. The *Erica carnea* yields great masses of plant residues which are difficult to decompose. They form humus layers in the surface soil somewhat similar in appearance to raw humus layers. But in spite of that similarity in appearance, the humus formation is different. This is evidenced by its different behavior. Raw humus layers are always unfavorable to forest vegetation, offer little chance for accrescence and are a poor bed for seed propagation. The humus layer below *Erica carnea* on limestone, on the contrary, is favorable to forest vegetation, and so propagation is not checked. By micromorphological investigation, the differentiation from raw humus (already evidenced by the profile morphology) becomes clearly visible.

We therefore designate this soil as Tangel-Rendzina, and the humus form as tangel humus. The development of the humus form is caused primarily by the plant cover *Erica carnea*. This plant, like other plants which have a similar effect on soil development, we call tangel plants. Other tangel plants are *Rhododendron hirsutum*, the hairy mountain rose, *Juniperus montana*, the alpine juniper, *Sarothamnus purgans*, the Spanish brook. "Tangel" is taken from an Austrian peasant term for the kind of leaf residues accumulated in the surface layers (mostly needle-shaped). The tangel plants are humus plants, that is, they develop in previously formed humus layers. In the case of our study, the prolific growth takes place in a mull-like Rendzina or in a Mull-Rendzina.

The Tangel-Rendzina in the Vienna Woods has the following profile characteristics: the litter (A_0 horizon) is much thicker than in rendzina moder, although not so thick as in the mountain and subalpine varieties. The F layers have a thickness of about 15 cm., the H layers contain calcareous mull consisting entirely of earthworm casts and are about 10 cm. thick. The calcium layer, rich in recrystallized $CaCO_3$ in the form of very small powdery crystals, is well developed and grades into the Ca/C layer, which still has a large amount of recrystallized $CaCO_3$ and a partly preserved rock structure. The C layer is fresh limestone or calcareous dolomite (Figure 5B, Chapter VIII).

V

Soil on Limestone in the Vienna Woods:
The A(B)C Stages

A complete soil development on limestone is produced in two phases: (1) in the phase of the Rendzinas (AC phase); and (2) in the phase of Terrae calcis (ABC phase). The contrast between the two phases is so great that one can hardly believe they have developed out of the same parent material.

To the naked eye the ABC soils on limestone in the temperate zone look very much like tropical or subtropical soils, and this impression is fully borne out by their micromorphology when they are investigated by thin section. The yellowish soil has the fabric of a tropical or subtropical Braunlehm, the red soil the fabric of a rubified Braunlehm, a Rotlehm or a Roterde. Their fabrics are very similar to those of the red and yellow podzolics in the United States, but where these limestone soils are found in the United States, they are not called red and yellow podzolics. One of the ABC soils bears one of the oldest names in soil science: Terra rossa.

If thin section analysis shows that Terra rossa is a soil formed by rubi-fication, then, according to our comparative micromorphological research findings, we must assume that it developed out of a braunlehm fabric. And this is actually the case. In the southern Vienna Woods the red variety, Terra rossa, does not occur; the yellow, ocher or brownish one with braunlehm fabric does. In central and northern Europe the yellow to brownish variety prevails and the red variety is rare. In the Mediterranean countries the Terra rossa predominates; the yellow variety is limited to special environments where development is typical for habitats with high humidity and no dry seasons. As a consequence of these findings, a name had to be found for the yellow to brownish variety—which had not yet been named or described in the literature. Since Terra rossa is a red soil, the brownish to yellowish soil on limestone with braunlehm fabric was named Terra fusca (fuscus = tawny).

As for the Terra rossa, we know that a rubified soil, in its further development in environments with marked dry seasons, increases in spaces, turns into a loose, water-stable fabric and becomes earthy. We named this process *rote Vererdung* (red earthening). Since this same process also occurs with limestone soils in a very typical way, we decided to refer to the earthy variety, corresponding to a tropical Roterde, as earthy Terra rossa. (It was formerly designated as mediterranean Roterde.) A further differentiation between the dense rotlehm-like Terra rossa and the earthy Terra rossa has not been established, though knowledge of it is essential to studies in soil development, environmental research and practical soil science.

How do we explain the fact that soils with the fabric of a tropical or subtropical soil are so common on limestone in strictly temperate climates, whereas they do not occur on silicate rock in the same environments? The answer can only be arrived at by comparative micromorphological research. I emphasize this and repeat it, because the thin section method of analysis offers unparalleled possibilities for the study of soils.

One of the characteristics of soil development on limestone is that the first phase, the Rendzina phase of development, is determined by the limestone as a parent material. The Rendzina phase is particularly well represented in the humid temperate climate. In no other climate can so many forms of it be found, or with so many different, well-characterized stages of development. It is not an accident that Rendzina research originated in the humid limestone regions of Poland, from which it gets its name.

THE MODES OF OCCURRENCE

As to the second phase, the ABC phase, we must keep in mind that one and the same soil type may be found in nature in six different forms or, simply, in six different modes.

If we select a particular soil type,* say Rotlehm from basalt, we can observe it in the following modes: (1) recent soil in situ, (2) relict soil in situ, (3) fossil soil in situ, (4) recent soil sediment, (5) relictic soil sediment, (6) fossil soil sediment (Figure 2). Sediments may be fluviatile, lacustrine, marine or aeolian. Whether it is a soil sediment (produced by soil formation before erosion and deposited as a sediment layer) or a sediment transformed by diagenesis (in a buried state after deposition) can be determined by thin section investigation.

---

* In the sense of Dokuchaev.

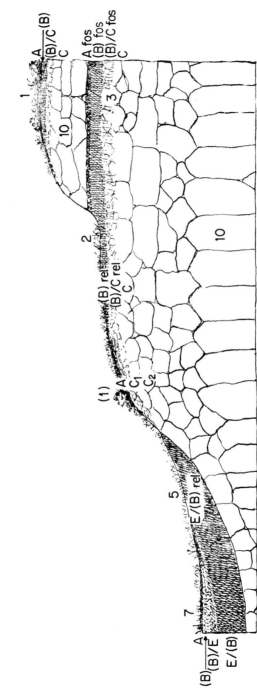

Figure 2. Paleopedological Modi of a Rotlehm Formed from Basalt

Figure 2 (cont'd)

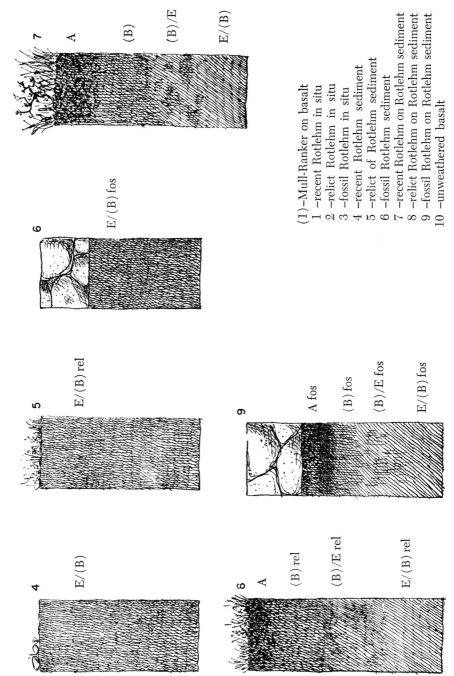

(1) – Mull-Ranker on basalt
1 – recent Rotlehm in situ
2 – relict Rotlehm in situ
3 – fossil Rotlehm in situ
4 – recent Rotlehm sediment
5 – relict of Rotlehm sediment
6 – fossil Rotlehm sediment
7 – recent Rotlehm on Rotlehm sediment
8 – relict Rotlehm on Rotlehm sediment
9 – fossil Rotlehm on Rotlehm sediment
10 – unweathered basalt

29

Since the fabric of Braunlehm, Rotlehm and Roterde is, in general, almost completely restored in soil sediments, it is sometimes difficult to distinguish between a (B) horizon material of a soil in situ and a sample of the corresponding soil sediment. There are many sediments that show no alteration; for example, marine sediments after deposition. A decisive alteration can be produced by reduction. The only cause of reduction on a large scale seems to be the influence of organic life or certain organic substances. Subaqueous humus forms which produce intense reduction in soil sediments are sapropel, sapropelitic gyttja and acid peat formations. Without the presence of reducing life or reducing organic substances, Braunlehm, Rotlehm and Roterde sediments remain almost unchanged. They are of great interest to the paleopedologist because they permit a theoretical reconstruction of soil covers that have been completely eroded and therefore no longer exist in their original environments. They have been replaced by soils that were formed by recent environmental conditions.*

Huge deposits of sediments of terrestrial soils are found in the shelf regions surrounding the continents. In the tropics, the sediments transported by the rivers into the sea are for the most part soil sediments. These sediments often give the rivers their yellow or reddish color. The color may persist twenty miles or more and even into the sea. The best-preserved examples are soil sediments imbedded in limestone. Sedimentary limestones are produced mainly by calcium carbonate-excreting animals in the sea. Large numbers of these animals developed in tropical seas, in pre-Quaternary geological periods.

How can soil sediments imbedded in limestone rocks be analyzed? By thin section analysis of the rocks if the sediments are present in high concentration. Or, where the concentration is low, by careful solution of the rock with diluted acids (best for this is acetic acid). The accumulated insoluble constituents can be used for thin section preparation, x-ray investigation and analysis with the electron microscope. In dissolving the calcium carbonate (with carbonic acid), nature follows a similar course. The insoluble constituents of the limestone rock are set free and deposited locally in spots that are especially suitable for their accumulation. We find them in fissures and caves, where they are known to the speleologist

---

* Compare relics in situ and sedimented in the form of rañas (mud and rubble streams in dry climates) in Spain (Chapter XXII).

as cave loam (*Höhlenlehm*). On free or open land they accumulate in depressions of the surface relief (dolinas of the karst landscape) or slope steps (breaks), slope ends, areas next to the slope ends and old plateau areas.

These areas are also the habitats of the ABC stages of soil sequence development on limestone. It is clear therefore that whereas the Rendzina stages have developed from the limestone as a whole through its function as a parent rock, the parent material of the ABC stages consists of only a small portion of the rock, the insoluble constituents; the main substance, the calcium carbonate, has been removed from it by solution.

Why show limestones primarily as remnants of tropical and subtropical soils? Because temperate and cool climates in our latitudes were historically very rare. Within the period of 350 million years in which terrestrial soil formation is assumed to have been going on, only about 1 million years can be cited for the formation of soils which correspond to some temperate or cool climate. That is less than 0.3 per cent. Also, outside limestone areas, soil relicts of geological periods prior to the Quaternary have, in general, the character of tropical or subtropical varieties. Furthermore, marine limestones are produced by the action of calcium carbonate-excreting animals, and these occur primarily in the seas of the tropics. Here the soils are highly subject to erosion, and this is why great quantities of erosion material of terrestric soils were transported to the shelf regions.

Going back now to our development studies in the southern Vienna Woods, we must begin with the stage of the Mull-Rendzina. (A further stage of the Rendzina development, the Tangel-Rendzina, must be regarded as an exceptional case in the Vienna Woods which cannot be included in the normal development sequence.) The Mull-Rendzina is characterized by a certain clay content which permits better water conditions, better development of the soil life, better structure (sponge structure) and aggregate formation.

The clay derives from the limestone; therefore, the habitats are those in which a certain accumulation of it can take place. A normal amount of clay has a favorable effect on the soil as a whole, in which case the soil still has the character of a formation typical of a temperate climate. Biologically, it is the most valuable stage of development, so that in Central Europe one speaks of the "Chernozem of the woods." If the clay content increases, the favorable properties diminish. The soil layer is thinner, as is the humus horizon, and the structure is more dense. The profile has a brownish seam on the lower brim of the A horizon. Because of the increasing content of free iron hydroxide, the A horizon also acquires a brownish tint. The thin (B) horizon may show characteristics of braunlehm fabric in a thin section. We call this soil a brown Rendzina.

TERRA FUSCA

In habitats where the Braunlehm sediments set free by solution from the limestone accumulate to any great extent, we observe very dense, ocher to brownish-ocher or reddish-brown lime-free deposits. The soils formed in these habitats have a much higher clay content than the brown Rendzina (which can be regarded as a transition stage). We call this soil a Terra fusca, that is, a soil similar to a Terra rossa but with a braunlehm fabric instead of a rotlehm fabric.

If the insoluble constituents in the limestones are primarily Rotlehm remnants, then a Terra rossa can be formed, in much the same way as a Terra fusca. This is only one of the ways for a Terra rossa to develop, however. There are others that will be treated later on.

Typical Terra fusca is a soil of a summer-humid temperate climate, a climate in which the Braunlehm characteristics remain preserved and are unaltered to any great extent. It is primarily a forest soil.

In the southern Vienna Woods the final stage or climax formation of soil development that is possible under the same environmental conditions is a Terra fusca. Even though it is in a Terra rossa sequence, ending with a climax formation with very advanced development, the initial stages are rather indistinct. The Rendzina-Terra fusca sequence appears the most suitable standard for comparative investigations with soils derived under other environmental conditions.

Humid Terra fusca is more or less free of lime. A Ca horizon as a rule is absent. The soil lies directly on the hard limestone rock, which exhibits a washed, smooth surface; however, some of the grooves and dents of karren that had been produced by solution of the calcium carbonate are now smoothed. In soils with recalcification, a secondary lime accumulation may occur in the transition layer to the underlying parent rock. The fabric in the (B) horizon is most representative and has a groundmass colored homogeneously yellow (egg yellow) by peptized amorphous iron hydroxide; this is movable by diffusion and can be precipitated in the form of round, dark-brown concretions with smooth surfaces, as in tropical Braunlehms. The structure is very unstable; aggregates (blocky joints and earthworm casts) are easily destroyed and silted up by wetting. The groundmass shrinks considerably with drying. But the shrinkage cracks close up again after some rain. The humus horizon is very thin (4 to 8 cm.), and is low in humus content. In the forest it exhibits mull in the lower part, and moder development in the F horizon. Almost all soil life is concentrated in the thin F horizon. Only some earthworms go deeper into the soil. In the southern Vienna Woods illite is the principal clay mineral, indicating that a more subtropical variety of the original Braunlehm is imbedded in the limestone. The weathering of the minerals is strong. In the small sand fraction, of which there may be almost none, quartz is practically the only mineral that has been left.

The development of a Terra fusca is a complicated and slow process. Remnants of a former soil of a humid tropical or subtropical climate formed in the Tertiary or before, destroyed by erosion, deposited on sea shelves, and imbedded in layers of calcium carbonate excreted by sea animals have become insoluble constituents of a limestock rock. They are set free by solution under the present environmental conditions and are accumulated into new Braunlehm layers. We call this process a restitution, and the soil layer restored a Braunlehm sediment.

From this example we can see that two different kinds of soil genesis can be distinguished:

(1) A monogenetic soil development: when a sequence of development stages is formed from the same parent material, in the same habitat and under more or less the same environmental conditions.

(2) A polygenetic soil development: when a sequence of development stages, beginning with the most simple and ending with the most complicated, is formed from the same parent material but under different environmental conditions, and in many cases in different habitats.

The Rendzina phase of soil development on limestone in the southern Vienna Woods is an example of a monogenetic soil development.

The brown Rendzina-Terra fusca phase in the southern Vienna Woods, which in other environments may be completed by one or two Terra rossa stages, is an example of a polygenetic soil development.

VI

The Terra Rossa Sequence of Soil Development

DO THE AC AND THE ABC STAGES BELONG TOGETHER?

The Rendzina phase (AC phase) of soil development is produced from limestone as parent material, and the Terra fusca or Terra rossa phase (ABC phase) is formed from the insoluble constituents of the rock—that is, from only a very small part of the limestone (elements that accumulated and serve as new parent material in certain habitats).

Why do we regard these phases as belonging together and not as distinctly separate?

(1) The two phases are not separate. Between the two most representative development stages, the Mull-Rendzina of the AC sequence and the Terra fusca of the ABC sequence, there is a series of clearly established transitional stages. Even the Mull-Rendzina has characteristics that anticipate the further development—primarily the increasing clay content. This differentiates it (as a Rendzina with mull formation) from the mull-like Rendzina and the Proto-Rendzina.

(2) Terra fusca remains a soil on limestone. The parent rock from which it is derived continues to represent an important horizon of the soil profile, and it greatly influences the character of the soil. Terra fusca is a humid soil, but it rarely displays gley formation.

(3) Normally, soil development proceeds from AC initial stages to ABC final stages; this is true on silicate rocks as well as on other parent materials. The AC stages develop relatively rapidly, whereas the development of the ABC stages is very slow. With soil development on limestone, the final stages are especially slow. In spite of this, each stage exhibits its definite connection to the previous stage and to the stage that follows. The relationship is proved by the fact that there are properties in each stage which anticipate the next stage, or which remain from the previous stage. We designate the first type as progressive properties, the second as retardatory properties. Acquaintance with them is important because

34

they reveal the exact position of a form in the natural development sequence. For example, in the southern Vienna Woods, a mull-like Rendzina and a Terra fusca occur within the same area; they are so different that no relationship at all could be found between the two. But as soon as we discovered the transitional stages, that is, the Mull-Rendzina and the brown Rendzina, it was apparent that both are members of a definite development sequence and we could recognize their exact position in it.

In the development to a Terra fusca in the Mull-Rendzina are the following progressive properties: the increase in clay content, the decrease in calcium carbonate and the decrease in microskeleton. Retardatory properties are the high lime content, the highly developed Ca horizon with recrystallized $CaCO_3$ and the presence of a certain amount of clastic calcite fragments.

In the brown Rendzina, the progressive properties are the decrease in lime content, the increase in accumulation of clay and free iron hydroxide, the brownish color, the beginning formation of a (B) horizon and the formation of braunlehm fabric in this horizon with distinct double refraction by particle arrangement in the groundmass. The retardatory properties are the relatively deep humus horizon, the rendzina-like character, the increase in humus content and the presence of a well-developed Ca horizon.

(4) A complete soil development on limestone (one which passes through all stages of the development sequence until the stage of the Terra fusca is reached) can be observed only in environments with conditions that allow a complete development. If the climate is too cold or too dry, only the first Rendzina stages can be formed. Intermediate climates allow the formation of a well-developed Mull-Rendzina; and sufficiently humid climates and forest vegetation make the final stages possible. This progression indicates that both phases, the AC and the ABC or Terra fusca phase, are parts of a unit. Only by consideration of the complete development sequence, and of the different ways in which the sequence could have been shortened because of some insufficiency in the complex of environmental factors, can a region be fully understood in regard to its pedological character. The natural laws governing the shortening of the development sequences can be represented by the scheme in Table 1.

Table 1 Rules Governing Shortening or Interference with Soil Development Series on Limestone as a Result of Climatic Conditions

Development Stage	*Limitations Caused by Climatic Insufficiencies*			
	Development reaches biological peak	Development reaches A(B)C stages	AC stages destroyed or transformed into Xero-Rendzina	AC stages, of rapid and indistinct development; unstable and easily eroded
1. limestone raw soil, 2. Proto-Rendzina, 3. mull-like Rendzina	Climate too cold or too dry and cold; Development limited to initial stages			AC stages, of rapid and indistinct development; unstable and easily eroded
4. Mull-Rendzina	Intermediate humid conditions; fairly warm and humid summers			
5. brown Rendzina		Humid temperate climate with forest vegetation		
6. typical Terra fusca				
7. Terra rossa		Humid mediterranean to subtropical climate with alternating hot dry seasons (now rare)		
8. recalcified Terra fusca and Terra rossa with xeromorphous changes			Secondary change of the above conditions, producing a much drier climate (now prevalent)	
tropical Terra fusca	Very humid tropical climate without effective dry season			Under the soil cover, the C horizon of the Terra fusca limestone area is frequently silicified and calcium carbonate has been removed by solution

EVALUATION OF THE DEVELOPMENT STAGES

Development is the gradual transformation of a natural body, beginning with some primitive stages and progressing to more highly organized ones. The different stages of soil development vary greatly in their biological value, which in essence corresponds to their value for agriculture and forestry. The more highly organized, the more valuable, in general, is the soil; however, transformation may produce a retrogressive development in some cases. The final stages of a soil may diminish its biological value. This has often been designated as the "aging" of the soil. There are soils which never age; there are others which are very susceptible to aging. It depends on their kind of development. In the case of soil devel-

Figure 3. Four Soil Development Sequences on Limestone and the Biological Values of the Different Stages

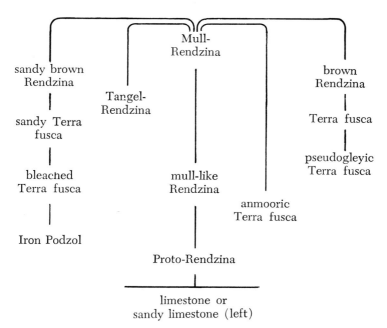

Note: The more distant the position of a development stage from the base, the higher its biological value. In the initial stages we see an increase in biological efficiency; in the final stages a decrease.

opment on limestone, we observe a marked decrease in the biological value in the last stages (see Figure 3).

Figure 3 shows, in addition to the genetic succession of the four development sequences, the biological values of the different stages. The more distant the stage is situated from the base, the higher its value. The three final stages are considerably less valuable than the Mull-Rendzina. The rare case of a sandy Terra fusca leads to a heavy leaching and the formation of a whitish, bleached sand layer as an A_e (eluvial) horizon. The anmooric and pseudogleyic Terra fusca can also be considered rare formations.

DEVELOPMENT OF TERRA ROSSA

This, to the layman the best-known name in soils, refers to a soil that forms an essential part of the Mediterranean landscape. It has one of the oldest of soil names. The Latin substantive was used by the Romans (*terra sigillata*, the red clay used for the bright-red Roman pottery), and Roman designations for some soil materials were used in the Middle Ages: *terra miraculosa saxonia, terra alba, bolus alba* (a white clay loam suitable for medical use).

Linné tried to introduce Latin names for soils, but he was not successful because soils were not sufficiently understood even in the eighteenth century, particularly their morphology. Today, Latin has ceased to be the language of soil science, and the traditional nomenclature has followed other ways.

Terra rossa is bright cinnabar to brick red, in some cases reddish cinnamon. The soil is, for the most part, poor in humus, decalcified and generally without formation of a Ca horizon if situated on the fresh limestone. The classical area of distribution is the Mediterranean karst landscape of Istria and Dalmatia.

This landscape characteristically has extremely poor vegetation and soil. In general, little more than bare rock surfaces are in evidence. Because of the heavy soil erosion, soil is found only in the cracks and grooves (karren) of the limestone. The only habitats with deeper accumulations of soil are the dolinas, the "red eyes" of the karst landscape; only there can the soil be used for grain crops. But since the "eyes" are too small for plow and harrow, the hoe is the only implement used for cultivation. Generally, the owner builds a small wall of limestone around his small, circular field. Because of the severe scarcity of soil in this country, theft of soil occurs repeatedly.

So many explanations of the genesis of Terra rossa have been given (some contradictory to each other) that the literature is very extensive, going back to the first decades of pedology. Thus, Terra rossa has been variously explained as a decomposition product of volcanic ash; as an iron oxide-enriched marine sediment of a particular sea, the Ferretto Sea; as aeolian deposits that originated in the African deserts; as a transformation product of a clayish marine mud rich in iron sulfide; as residues of a lateritic soil formation; and as a recent soil formation produced by the present Mediterranean climate.

Some of these explanations recognized the presence of accumulations of insoluble constituents of limestone. On the question of why the percentage of iron is much higher in many Terra rossa varieties than the sum of insoluble constituents in the limestone, almost as many contradictory explanations arose: for instance, iron-rich solutions, which are abundant in the karst landscapes, enriched the clayish deposits; the cause of iron enrichment is the limestone itself, since dissolved iron is precipitated in contact with calcium carbonate; the iron enrichment was produced by metasomatic displacement of the calcium carbonate by iron solutions. Another theory explained the enrichment as having been produced by capillary rise of iron-enriched solutions to the rock surface, as a consequence of the heating of the limestone in the hot Mediterranean summer climate. According to another explanation, the iron, on the contrary, was moved downward, forming a reddish illuvial horizon (the iron having been protected by acid humus soils produced in the humus horizons of former forest soils which had been completely eroded after the climatic changes).

I have a completely different explanation, one which may seem rather bold after so many other opinions; however, I was able to use methods that gave a much clearer picture of the facts. For this explanation one needs only to be able to recognize a braunlehm fabric, to know the processes of its transformation, and to know the fabrics of the transformation products. My discoveries were made in the course of continuous comparative investigations of tropical soils of the Braunlehm, Rotlehm and Roterde groups. The processes involved in the corresponding transformations of braunlehm fabrics are rubification and red earthening (*rote*

Vererdung). The final transformation stage is always accompanied by a relative enrichment of iron by the leaching of the silica-rich, nonprecipitated residues of the former groundmass. It will be recalled that rubification and *rote Vererdung* are processes opposite to laterization.

Terra rossa occurs not only in the Mediterranean countries of Europe but also in all countries with a mediterranean climate all over the world. However, in most cases it represents relict soils. The countries of the Mediterranean littoral have become too dry for it to form. In more humid and cooler latitudes, Terra fusca is much more common; however, some isolated areas of Terra rossa are found in temperate regions, for example, in the limestone Alps, in Moravia, in Bavaria and even in Britain, where Terra rossa occurs in Somerset and North Wales. To explain that phenomenon the genesis of the two forms of Terra rossa must be clear: the allochthonous Terra rossa and the autochthonous Terra rossa.

The allochthonous Terra rossa is produced on allochthonous Rotlehm or Roterde sediments, on and from limestone rubified elsewhere in earlier tropical or subtropical habitats.

The autochthonous Terra rossa has its genesis in subtropical or mediterranean regions. The effect of the dry season is much intensified by the underlying limestone. Of particular influence is a humid, wet season. The allochthonous Terra rossa may occur in cooler climates if enough solution of the limestone takes place and a liberation and accumulation of its less soluble constituents are produced.

The two genetic varieties can be distinguished by comparative thin section analysis of the rock and the soil.

VII

Alpine Soils on Limestone: General

Since the Austrian Alps are composed in large part of high limestone mountains, our investigations here were susceptible of comparison with those obtained in the southern Vienna Woods and other lowland and hilly limestone sections. The differences are very characteristic, and have led to distinguishing several alpine varieties of Rendzinas. Their formation is the result of the very high mountain climate, especially that of the highest zones, the nival and the alpine zone.

Temperatures in the belts above the timberline and krummholz line are much lower than in the corresponding lowlands. The nival zone is like the high Arctic, a cold desert. In spite of the abundant precipitation, environmental conditions are barely favorable for the development of life. Even in the adjacent grassland belt of the alpine zone, environmental conditions support only a very specialized type of life.

At low mountain temperatures, it snows more frequently and abundantly, and the snow remains longer on the ground than is the case in the lowlands. In the mountains there are areas in which snow does not melt at all. Generally, however, heavy snowfalls that occur in summer soon melt away. I myself ascended to an altitude of 3,000 m. in August in such a heavy snowfall that I gradually got stuck in a snow cover of 75 cm. I spent the night in the open, far from a mountain hut, and I had to be searched for the next day.

Of great importance to soil development is the fact that the soil is frozen under the snow cover most of the year; only in summer does the soil thaw. In spite of the fact that it sometimes snows heavily in summer, the difference between winter and summer is tremendous.

This is in contrast to the nival and grassland belts of the high mountains in the tropics, where there is little seasonal variation. Above the snow line, it is winter the year round, whereas in the grassland belt, for example, at the altitude of Quito (2,850 m.), the attractive capital of Ecuador, it is more or less perpetual spring.

In the alpine grassland belt of the Austrian Alps the summer is very marked. Although the nights may be cold, the sunny days get warm. A person feels warm only when he sits or moves directly in the sunshine, for in the shade the air stays cool. Yet, the air and sunshine are very agreeable. The air is clear and bracing, and the sun has a stimulating effect on life. The vegetation is dense but low. An abundance of alpine flowers with brightly colored blossoms contributes to the splendor of the alpine landscapes.

Hot sun, cool shade, high precipitation, frequent snowfall, long-lasting snow cover, thawing of frozen soil in spring (which in the frozen state is impermeable)—all promote the development of characteristic soil varieties. But first, what are their habitats, their vegetation and the altitudes at which they occur? The plant cover, soils and life forms vary in different belts according to altitude and also according to latitude and longitude.

TANGEL-RENDZINA

In discussing soil development on limestone in earlier chapters, I referred to a Rendzina-Terra fusca sequence typical for the hill belt and the mountain belt up to an altitude of about 600 m.; however, near the upper limit of the mountain belt in the southern Vienna Woods, there occurs a Rendzina variety called Tangel-Rendzina. It was excluded from our standard scheme as being extraordinary, not normal for the soil development of the southern Vienna Woods. Whereas the other Rendzinas tended to develop toward Terra fusca as a final stage, and displayed the characteristic progressive properties in their transition stages for such a development, the Tangel-Rendzina showed a different tendency. In fact rarely does a well-developed Tangel-Rendzina become transformed into a Terra fusca.

Tangel-Rendzina is a typical mountain soil, and we know now that this variety is the chief soil formation on limestone of the lower and upper subalpine belt as well as of the krummholz belt. Here there is something that accounts for all alpine Rendzinas. Every one of those listed in Table 2 represents a final stage. They are climax formations in spite of the fact that they belong only to the AC phase. They have no property whatever that can be regarded as a property progressing towards an ABC development.

Table 2   Hypsometric Change of Soil Profile Development in the Limestone Alps of Northern Austria

Climax Formation	*Phases of Profile Development*
8. alpine carbonate Råmark of the nival zone (above 2,800 m. = 9,200 ft.)	(A)C
7. alpine Cushion-Rendzina, belt of the alpine cushion plants (up to 2,800 m. = 9,200 ft.)	(A)C, AC
6. alpine Pitch-Rendzina of the alpine grassland belt (up to 2,700 m. = 8,850 ft.)	(A)C, AC
5. Tangel-Rendzina of the dwarf shrub belt, vegetation *Pinus mugho* rich in mountain roses (up to 2,400 m. = 7,900 ft.)	[(A)C], AC, $A_F A_H C$
4. Tangel-Rendzina of the spruce forest belt, vegetation Ericeta-Piceeta (up to 2,000 m. = 6,600 ft.)	[(A)C], AC, $A_F A_H C$
3. Tangel-Rendzina of the pine forest belt, vegetation Ericeta-Pineta (up to 1,800 m. = 5,900 ft.)	[(A)C], AC, $A_F A_H C$
2. mull-like Rendzina and Terra fusca, vegetation pine and deciduous forests (up to 1,000 m. = 3,300 ft.)	[(A)C], AC, A(B)C
1. mull-like Rendzina, Mull-Rendzina, Terra fusca, vegetation pine and deciduous forests, agricultural crops (up to 600 m. = 1,970 ft.)	[(A)C], AC, $AC_1 C_2$, A(B)C

TERRA FUSCA RELICTS IN THE ALPINE GRASSLAND BELT

Still, Terra fusca varieties do occur in some environments of the limestone Alps (Figure 4). These Terra fusca varieties tell another story. They contain well-developed iron concretions, but in a completely transformed earthy groundmass as a result of the mountain climate. The presence of Terra fusca above the timberline, in an area where soils evidently do not

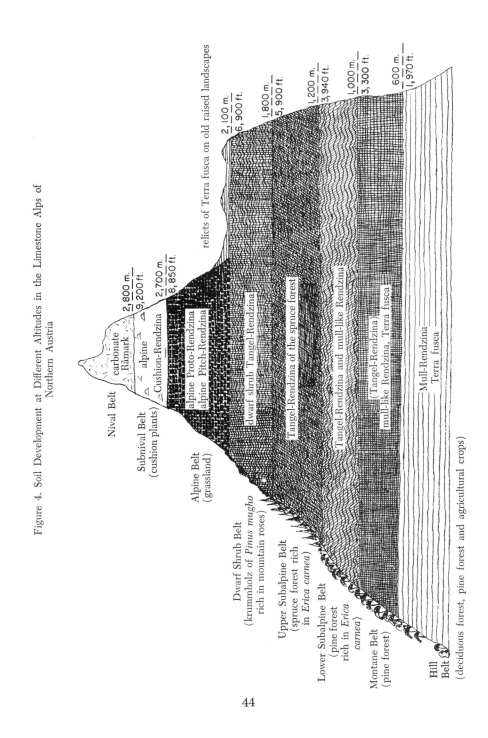

Figure 4. Soil Development at Different Altitudes in the Limestone Alps of Northern Austria

relicts of Terra fusca on old raised landscapes

2,800 m.
9,200 ft.

2,700 m.
8,850 ft.

2,100 m.
6,900 ft.

1,800 m.
5,900 ft.

1,200 m.
3,940 ft.

1,000 m.
3,300 ft.

600 m.
1,970 ft.

Nival Belt

Subnival Belt
(cushion plants)

Alpine Belt
(grassland)

Dwarf Shrub Belt
(krummholz of *Pinus mugho*
rich in mountain roses)

Upper Subalpine Belt
(spruce forest rich
in *Erica carnea*)

Lower Subalpine Belt
(pine forest
rich in *Erica
carnea*)

Montane Belt
(pine forest)

Hill
Belt

(deciduous forest, pine forest and agricultural crops)

carbonate
Råmark

alpine
Cushion-Rendzina

alpine Proto-Rendzina
alpine Pitch-Rendzina

dwarf shrub Tangel-Rendzina

Tangel-Rendzina of the spruce forest

Tangel-Rendzina and mull-like Rendzina

(Tangel-Rendzina)
mull-like Rendzina, Terra fusca

Mull-Rendzina
Terra fusca

44

develop beyond the AC stages, must be regarded as unusual. Since in the interglacial periods conditions similar to those of present times were effective, one must assume that the Terra fuscas within the alpine zone represent relict formations of the Tertiary period. With the geomorphologists (Wiche, 1951), I am convinced that these relicts of Terra fusca are part of old Tertiary landscapes that were raised to a high altitude by the upward thrust of the mountain formation. With this vaulting, evidently the entire landscape with its hilly topography, including the soil cover, was raised. The Terra fusca was eroded a great deal, and the residues were transformed under the influence of the present alpine environmental conditions (Plate 3C and 3D).

ISOLATION OF THE OLD LANDSCAPES

In the Quaternary period, with its climatic changes and the great glaciers of the ice ages, with its deep-cutting valley formation, and with the disappearance of streams from the surface and the karst formation, the old Tertiary landscapes were isolated and their soil development was arrested.

From the residues and the micromorphology of the soil remnants, the former conditions can be reconstructed. The warm, humid climate and the intense development of life must have brought about heavy weathering and a high degree of solution which created conditions similar to those in the humid tropics or subtropics of the present: virtually no Rendzinas[*] and no karst formation, the silting up of Terra fusca material and the filling of all cracks and fissures of the rock with clay and loam sediments.

Terra fusca is present only in remnants that may occur in connection with newly formed Rendzinas; it is mixed in these or lies separate, below the humus horizons. Sometimes even rubified remnants of Terra fusca can be found. Since nowhere on earth is rubification known to take place in an alpine environment, their nature as soil relicts is clearly established.

Former tropical or subtropical soils raised to alpine altitudes by mountain formation, and subsequently transformed on silicate by present environmental conditions, have also been found on silicate rock in other high mountains (in the Andes of Peru and in the Sierra de Guadarrama in Spain), and will be the subject of future investigations.

[*] In spite of my extensive travels, I have not been able to collect Rendzinas in the humid tropics.

THE ALPINE RENDZINAS

Going back now to the alpine Rendzina varieties, I shall try to give such information as is necessary for an understanding of the whole group, but without treating them all in detail.

The alpine Rendzinas develop from simpler initial stages, and continue into those stages which represent the climax formations in their environments. Tangel-Rendzina must begin with a Proto-Rendzina and have at least a mull-like Rendzina in which the humus plants of the tangel vegetation can find the proper substrate for prolific growth. Once formed, Tangel-Rendzina in the subalpine belt undergoes no further development. This is the case even with tangel soil varieties on north slopes of the subalpine belt and the dwarf shrub belt; these attain a character similar to that of a raw humus (dystrophic Tangel-Rendzina).

ALPINE PROTO-RENDZINA

The very characteristic Rendzina of the alpine belt, the alpine Pitch-Rendzina, begins with a Proto-Rendzina stage. This soil is different from the very shallow Proto-Rendzinas of the hill zone; it may be rather thick (up to 3 m., or 10 ft.) and consist of rock detritus (clastic calcite) and cylindrical droppings of small arthropods. Since the rock detritus (coarse and fine sand) is preponderant, the soil has a light-gray color, although the droppings under the microscope have a blackish color (Plate 2D). With further development, the rock detritus is dissolved and the amount of droppings increases. Finally we get a black, almost pitch-like organic soil in which almost no calcite grains are left (Plate 2E).

ALPINE PITCH-RENDZINA

The vegetation in this stage is generally a *Carex firma* grassland heath (a stiff, spiny sedge), a humid variety of the Firmetum. The profile of a well-developed Pitch-Rendzina has a depth of 30 to 35 cm., and it can be divided into an upper A_1 horizon (somewhat brownish in color because of its high content of plant roots) and a deep-black A_2 horizon. The A_1 horizon has almost no förna and no F horizon, which indicates a good humification. The Ca horizon, very characteristic for Tangel-Rendzina, is as a rule missing. The rock underneath has a smooth surface, though it is full of holes, grooves, and other karren formation as a result of the locally very effective solution weathering (the result of high humidity and intense activity of soil fauna).

The soil mass consists almost entirely of organic substance; it looks dense to the naked eye (though porous with the microscope), is fragile, and has mat surfaces at fractures but becomes smooth and shiny when exposed to rain and melting snow. It has a high water-suction and water-retention capacity. It seldom dries out. In occasional very dry summers, when all other soils dry out completely, the Pitch-Rendzina remains humid. On the other hand, water never stagnates, even in very humid periods. With loss of humidity the soil mass shrinks and has many fine shrinkage cracks. The shrinking, however, is not very severe.

If the soil mass consists primarily of small cylindrical droppings (sometimes other shapes also occur), they are usually 30 to 50 mm. in length. Their biology is not yet known, although it is assumed that they were produced mainly by collembola and enchytrae (Plate 2E). (The droppings are not loose, but stick to each other like boiled rice grains.)

Plant residues are humified to a great extent; the calcite grains have almost completely disappeared. If some calcite grains are left in the fabric, their forms are completely rounded by the intense solution weathering in the soil (Plate 2E). (Their forms remind one of melting ice cubes.) The Pitch-Rendzina is the most typical example of a humid Rendzina. The humus form is a kind of rendzina moder (pitch moder), but it has no moder smell (it is odorless). It is poor in bacteria and fungi, though there do exist some dystrophic soil varieties with a higher content of dark-brown fungus mycelium. They are also rich in fungus sclerotia, indicating that conditions for fungus growth sometimes become unfavorable. These soil varieties are much less humified, and frequently exhibit little-decomposed plant residues which have been reduced in size. They are transition forms to dystrophic coarse moder, which in lowland climates of the middle latitudes occurs only on acid silicate rocks, not on limestones.

In higher altitudes in the subnival zone, above 2,700 m. (8,850 ft.), the vegetation is concentrated in open patches within a more or less bare area. Most of these patches are very dense, small, rounded vegetation cushions which contain a very large number of minute individual plants. This is the best way for them to resist the unfavorable climatic conditions, which are counteracted only by the direct radiation of the sun on sunny days. In the interior of the cushions we find a development similar to soil formation in the case of the Pitch-Rendzinas of the alpine grassland heaths. It begins with a raw soil and continues over an alpine Proto-Rendzina to a kind of Pitch-Rendzina. The development is slower; thus the raw soil stage is prolonged. This is why cushion raw soils are numerous in this belt (Figure 4).

MOUNTAIN RÅMARK

In the nival zone, the raw soil becomes the climax formation. The development sequence consists of only one phase; the initial stage is also the final stage. (There are parallels in the dry desert, where unfavorable conditions are produced by lack of humidity.) Because of the unfavorable environmental conditions, no other soil formation is possible; not even the development of a humus horizon. A similar phenomenon occurs in the high Arctic. Both raw soils are formations of a cold desert. Both are known by the Swedish name Råmark and are distinguished by the subtype names arctic Råmark and mountain Råmark. Both are developed on and from limestone, and are therefore also classified as carbonate Råmarks.

VIII

Alpine Soils on Limestone: Special Variations

ABRIDGED SOIL DEVELOPMENT

In the Alps as well as in other high mountains in the middle latitudes a pronounced increase in precipitation occurs with increasing altitudes. The high mountains are high-precipitation islands. Abundant precipitation is necessary for effective solution weathering, and this latter condition creates landscapes in which Terra fusca is left as the only soil type of the soil development taking place, there being no associated Rendzinas. These landscapes occur in the humid tropics where the limestone is covered mainly with sediments of Terra fusca. Here the Terra fusca micromorphology presents a typical bolus-like braunlehm fabric, in some cases accompanied by signs of incipient laterization. The surface of the limestone rock is not only solidified but decalcified. It has the appearance of limestone, including its white color, but has actually been transformed into a dense chert which has no effervescence when treated with hydrochloric acid.

Why is very little or no Terra fusca found in the area of the Tangel-Rendzina within the subalpine belt of the *Erica*-rich spruce forests and the krummholz belt of *Pinus mugho* and *Rhododendron hirsutum* vegetation? The answer lies in a comparative study of a typical Tangel-Rendzina and a typical raw humus formation of a Podzol (Table 3; Figure 5).

TANGEL-RENDZINA AS A CLIMAX FORMATION

Tangel-Rendzina forms a very efficacious protective cover against erosion and dissolution above the limestone. Dissolved calcium carbonate is accumulated and reprecipitated in the strongly developed Ca horizon and Ca/C horizon. Calcite is brought back into the A_1 horizon and into the tangel layer by animal activity (particularly when dominated by earthworms). In addition to the protection by the Ca horizon, very effective protection is afforded by the mighty humus layer.

49

Table 3   Tangel-Rendzina and Podzol Raw Humus: Differences in General
          Properties

Raw Humus	Tangel Humus
Always very acid	Moderately acid in A_0, neutral in A_1
Contains acid humus sols	No formation of acid humus sols
Heavy leaching of colloid substances	No leaching of colloids
Produces bleached sand layers	No bleached sand layers
Humus layer with sharp border	Humus layer passes gradually into nonhumus layer
Very thin H layer	Preformed thick H layer tends to spread upward and downward
B_h layer	No B_h layer
No tendency to form mull-like moder	Tends to increase mull or mull-like moder layer
In forest zone, always unfavorable for forest growth	Favorable for forest growth; in H layer good decomposition and humification
Little decomposition, poor humification, little development of soil fauna, scanty animal droppings	Good development of soil fauna, rich in animal droppings of many kinds

The mull-like layer consists largely of water-stable aggregates, and the soil mass has a high water-holding capacity. It absorbs a large amount of rainwater, so that only a small part of the water reaches the Ca horizon and is absorbed in it. A considerable amount of rainfall is retained by the foliage of the vegetation, the pine or spruce trees and the dense *Erica, Rhododendron* and *Pinus mugho* shrubs. Much of the water is necessary for the growth of the forest and its shrubs and herbs. Another part is used up by the soil life. The humus layer was of special interest. This layer, consisting of A_0 and A_1 horizons, may reach a thickness of 50 to 100 cm. (20 to 40 inches).

How effectively it retains water can be observed in a distinctive soil of northern Spain, the Atlantic Ranker of Galicia. This soil is an AC soil, developed, not on limestone, but on granite. It occurs in a hilly landscape covered primarily with oak (*Quercus pyrenaica*), chestnut or pine

(*Pinus pinaster*) forests which have dense, dwarfed shrub vegetation of *Erica, Ulex, Genista* and *Sarothamnus* species. The climate is mild and humid (about 1,500 mm. or 60 inches of precipitation annually). The leading soil formation of the region is a humid Braunerde or a Semipodzol developed from it (podzolic Braunerde). Both soils have well-developed B and B/C horizons that have undergone rather strong chemical weathering. Within the area of ABC soils occur AC soils which have deep humus horizons but have not been subject to much chemical weathering and therefore retain many of their original characteristics. Since we call AC soils on silicate rocks Ranker, Franz (1956) designated these varieties as Atlantic Rankers (the humus formation being favored by the humid and mild sea climate of the Atlantic coast).

Here, too, the retardation of the development to the ABC stages is caused by the development of unusually thick humus horizons. These humus layers are influenced by the vegetation and by a very active earthworm fauna which produces a well-developed and water-stable sponge fabric high in pore space. Because of this fabric, the soil has an exceptionally large water-holding capacity. In a paper devoted to these soils José María Albareda (1964) reported that the water-holding capacity of such a humus horizon was 80 per cent of the soil volume. Given a thickness of 150 cm. for the humus layer, a rather continuous rainfall of 1,100 mm. per square meter would be needed to saturate it. Since the average precipitation of that area is 1,980 mm. for the entire year, the humus layer never does seem to become saturated. Furthermore, the soils dry out a great deal during the summer. It is understandable, therefore, that the subsoil receives very little of the rain, in spite of its abundance. This explains the low degree of weathering in the subsoil, as a result of which a large part not only of the potassium feldspars but even of the biotites remains almost unchanged.

As for the belt of the alpine Pitch-Rendzina, no soil development beyond the AC phase is apparent. The Pitch-Rendzina is a climax formation. But why is an ABC stage in the form of a Terra fusca not produced? The Pitch-Rendzina is subject to heavy leaching of the soluble constituents, particularly calcium carbonate (without leaching of colloid substances). In the case of the Tangel-Rendzina, dissolved and leached calcium carbonate accumulates in the Ca and Ca/C horizons, which horizons are usually missing in the profile of the Pitch-Rendzina. The Pitch-Rendzina, morever, never develops a humus horizon as thick as that of

Figure 5. The Biology and Micromorphology of Three Forest Soil Profiles with
Deep Humus Horizons

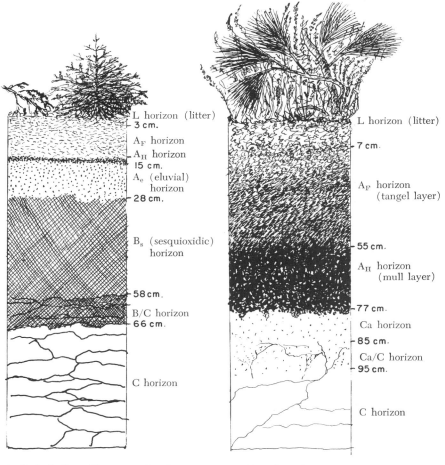

L horizon (litter)
3 cm.
A_F horizon
A_H horizon
15 cm.
A_e (eluvial) horizon
28 cm.

B_s (sesquioxidic) horizon

58 cm.
B/C horizon
66 cm.

C horizon

L horizon (litter)
7 cm.

A_F horizon (tangel layer)

55 cm.
A_H horizon (mull layer)

77 cm.
Ca horizon
85 cm.
Ca/C horizon
95 cm.

C horizon

A. *Podzol Raw Humus under Spruce Forest on Gneiss, Austrian Alps*

B. *Tangel-Rendzina under Austrian Pine Forest on Limestone, Austrian Limestone Alps*

52

Figure 5 (cont'd)

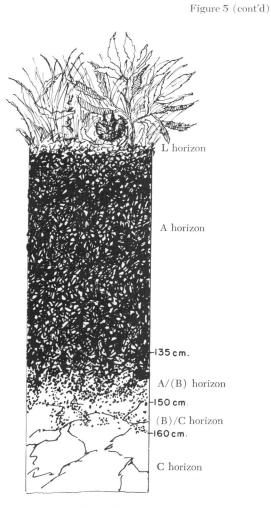

L horizon

A horizon

135 cm.

A/(B) horizon

150 cm.

(B)/C horizon

160 cm.

C horizon

 raw humus fabric

moder fabric

B$_s$ fabric

 tangel humus fabric

 mull fabric

 mortar fabric

braunerde fabric

C. *Brown Atlantic Ranker under Chestnut Forest on Granite, Northern Spain*

a Tangel-Rendzina. The rock surface below the Pitch-Rendzina profile
exhibits the effect of the strong dissolving action of the soil solution, so
that the surface is smooth even in the many grooves, hollows and channels
of the karren formation.

The micromorphology of the Pitch-Rendzina is the most typical for
the humid Rendzinas, dissolution of the clastic calcite in the humus layer
being almost complete (Plate 2E). How can the absence of ABC soils
in such an environment be explained, particularly in view of the high
precipitation? The answer lies in the fact that the precipitation falls
chiefly in the form of snow, and that the landscape is frozen for a large
part of the year. The effect of dissolution is intense locally but not in the
landscape as a whole. The melting snow has almost no dissolving effect.
The water must become enriched with carbon dioxide to be a solvent
for the carbonate which is produced by the vegetation and the soil life.

The most effective dissolution on limestone in the tropics and humid
subtropics is produced by heavy precipitation in the form of rain, by
luxuriant development of vegetation and life in general and by high
temperatures throughout the year.

THE DEVELOPMENT OF PARARENDZINA

In rocks that are rich in calcium silicates in the humid climates of the
temperate, subtropical and tropical zones, the calcium is set free by
weathering and disappears from the soil as a result of intense leaching,
at first from the upper layer and finally from the subsoil. In a climate
with marked dry seasons, the leaching is much less pronounced and in
some cases considerable amounts of calcium carbonate may even accumulate in the surface layers. The best example of such a development on
limestone is the formation of the lime crusts in Israel; other lime crust
soils or lime crusts (croûtes calcaires) on weathered rocks might also be
cited.

Lime crust formations in summer-dry climates occur also on certain
silicate rocks, particularly on basalt. The bare surfaces of the basalt rocks
in coastal zones are covered by a whitish coating that gradually develops
into a hard chalk layer that continues to grow in thickness. This layer is
produced by the weathering of the calcium-rich feldspars, primarily of
the anorthites of the basalt, with a composition of $CaAl_2Si_2O_8$. The calcium, by replacement, is transformed into calcium hydroxide, which is
changed by the presence of CO_2 into calcium carbonate. This also hap-

pens in humid climates, but the calcium carbonate is washed out continuously. In regions with dry seasons, it accumulates.

Lime crust soils, like lime crusts on rock surfaces, are formed primarily in coastal areas. Some humidity is indispensable for the weathering of the plagioclases and for the movement of the calcium solution by capillary rise. This humid condition is induced by the coastal climate, that is, primarily by high humidity and only to a small degree by rain. Without humidity, for example in the desert, no lime crusts can be established.

Enrichment by calcium carbonate as a result of the weathering of calcium silicates may also occur in semidry climates on soils without crust formations. This produces calcite-enriched AC soils and even calcite-enriched ABC soils. Of particular interest here are the AC soils. On these there develops a calcite-rich humus layer which in principle can hardly be regarded as different from a Rendzina derived from limestone rock (in this case a Xero-Rendzina). In spite of this, it does not correspond to the definition of a Rendzina, which is a humus formation on limestone (dolomite, magnesite or gypsum rock). Our parent material is a silicate rock. There are other examples with similar plagioclase-containing parent rocks, such as basanites, tephrites and even plagioclase-gneisses.

The soil formed on silicate- and carbonate-bearing rock was named a Pararendzina; however, this designation has a broad application. In the case of calcareous arkoses, sandstones with a marked feldspar content, for example a gneiss rich in plagioclases, might form a similar Rendzina-like soil. On the other hand, the Polish pedologist Miklaszewski (1922) separated Rendzinas developed from calcareous sandstones or shales from the limestone Rendzinas, and referred to them as mixed Rendzinas. There are other minor varieties of Rendzinas which form on crystalline schists, these being either calcium silicate rocks or calcareous mica schists rich in calcium carbonate. They must be distinguished from the limestone Rendzinas in humid climates for many reasons.

All the Rendzinas in humid climates represent only short initial stages, since they are subject to a rapid development to Mull-Rendzina or to ABC stages, for example, Braunerde varieties, Semipodzols or Iron Humus Podzols, which are climax formations. We call such Rendzinas Pararendzinas. Accordingly, a Pararendzina is a calcareous AC soil developed from and situated on calcareous silicate or siliceous rocks.

The rôle of calcareous silica and silicate rocks in the formation of Pararendzinas is very apparent on calcareous mica schists in the alpine grassland belt of the Austrian Alps. The principal constituents of the rock are quartz, muscovite, calcite and small pyrite grains. With this composition and the alpine environmental conditions, the development sequence is determined at the outset. Of special interest is the fact that this ABC sequence is found in a zone where, on limestone, only Rendzinas can be developed.

Soil development on calcareous mica schists in the alpine grassland belt parallels, in a very typical way, the development of the vegetation. Whereas soils on pure limestone develop strictly basophilic Dryadeta-Firmeta vegetation (grass heath of *Carex firma* with *Dryas octopetala*), the soils on calcareous mica schist very early develop vegetation of neutrophilic Sesleria-Sempervireta (grassland of *Sesleria sempervirens*) and Elyneta (*Elyna myosuroides*). In spite of this, the Proto-Rendzina in its first stage is produced in much the same way as it is on pure limestone. The first well-developed humus layer is formed under a trellis of *Globularia cordifolia*, just as it is in the southern Vienna Woods. The humus horizon attains a thickness of about 4 cm., is black in color, contains very small cylindrical droppings of mites and collembola and has a more or less neutral reaction. The parent material underneath is fresh, calcareous and whitish in color.

The rock, however, is soon decalcified, acquires a coarse-textured, sometimes rough and scurfy surface, becomes loose and brittle and changes its color from white to brown. In the disintegrating, crumbly groundmass the numerous sparkling mica plates are most conspicuously displayed. (These are called cat silver by the mountain people.) For mountain climbers, this half-disintegrated mica schist is very tricky rock because of the dangerous grips and stands. With the decalcification of the rock, the soil formation and the vegetation are completely changed. The Rendzina thus represents only a short beginning.

The transformation into Ranker is accompanied by much intensified humus formation. The humus attains a thickness of 10 to 13 cm., its pH drops to 5.5 or less. The forms of the droppings are much less well preserved in shape; in thin section there appear an abundance of bright-red decomposition products. The humus mass is more dense than in the Rendzina stage; it holds together and can be cut into peat-like bricks. After heavy rains or under melting snow water, it becomes soft, greasy, blackish in color and almost anmoor-like.

In habitats subject to strong winds, little decomposition of plant residues takes place. One finds peat-like, but strictly terrestric, humus formations which look somewhat like raw humus but whose genesis and character are very different; from the Russian terminology, they are known as eilag humus. The soil is called Eilag Ranker.

As soon as the fabric of the decalcified and weathered rock breaks down, the formation of a deep-brown B horizon over the weathered transition horizon (B/C horizon above the fresh parent rock) takes place. In fact, it is a (B) horizon; the soil has a loose braunerde fabric and is called alpine Sod Braunerde. It develops gradually to a kind of Iron Podzol, which, like all alpine Podzols, develops with reduced and irregularly shaped horizons and is therefore called alpine Sod Podzol.

IX

Soil Development on Silicate Rocks in Temperate Zones

INFLUENCE OF TEMPERATE CLIMATES

Soils develop in the greatest variety and undergo the greatest number of stages of development in regions with temperate climates. Within the temperate zones, the greatest variety of forms is attained in the humid temperate zones, although in the adjacent summer-dry steppe regions soil life activity and humus formation are extraordinarily intense. The diversity of processes surpasses even that of soil development in the tropics.

The tropics do possess a large number of transformation varieties of mature soil formations (for example, the many different transformations of Braunlehm); and therefore, numerous microscopic fabrics are produced by the different precipitation forms of iron hydroxide. However, what occur rarely or not at all in the tropics are the various initial stages of development. Since soil formations of previous geological periods in the present-day tropics were to a great extent similar to those of the present time in the temperate zones, the paleosoils of the tropics do not exist in as great variety as do the paleosoils of the temperate zones.

Thus, in regard to polygenesis, the temperate climates offer much greater possibilities for the formation of different soil varieties because both former and present climates have offered greater diversity. Pleistocene glaciation affected most regions that now have temperate climates and hence contributed not only to the shaping of the landscapes but also to the enrichment of the soils.

ENVIRONMENTAL CONDITIONS OF TEMPERATE CLIMATES

In examining the existing classifications of climates, to find concepts suitable for the zonation of soil development, it seemed advantageous not to separate the extreme winter-cold zones from the mild winter-cold zones. The regions with temperate climates extend, by this definition, from the regions of mediterranean soils up to the polar circle. Like the tropical regions, temperate regions are subject to considerable climatic deviations. All, however, have a certain characteristic in common which is critical for soil formation: they all have mild summers. The deviations are created by differences in the amount of precipitation and in the degree of temperature drop in winter. It is true that a climate of high continentality can scarcely be experienced as a temperate one, in view of the extremely cold winter. In spite of this, the Chernozem profiles just east of Vienna (and even the buried profiles inside the city) are very much like those in central Russia and in the district of Omsk in Siberia; and formations like the Iron Podzols of Finland are found in Austria.

WHY TEMPERATE CLIMATE?

The term "temperate" refers to a climate that typically yields the most agreeable and salubrious environmental conditions for man. It approaches the concept of a "comfort climate." The optimal comfort environment is considered to be one with a range of 60° to 70°F. (16.5° to 20.3°C.) in humid air, and 75° to 86°F. (22.9° to 29.8°C.) in dry air. This indicates that not only temperature but also humidity conditions are important, particularly when assessed for their influence on labor. The optima for physical work are temperatures of 68° to 86°F., with a relative humidity of 60 to 70 per cent; for mental work a temperature of 74°F., with a relative humidity of 55 per cent.

With respect to soil development, such climatic conditions are favorable for strong but not excessive chemical weathering (the second leads to a premature impoverishment in mineral nutrient reserves) and for an optimal development of the soil life, that is, a great variety of species with a high and many-sided biological activity. The result is optimal decomposition and humification of the organic residues, which, given adequate organic parent materials, lead to the creation of the best and most effective humus formations of the earth.

Inhabitants of the temperate zones like to look to the south and consider the Mediterranean climate or certain tropical climates as coming closest to the idea of a comfort climate. However, those who have lived for a long time in such climates know that the summers are not very agreeable. They are generally too hot and tend to be excessively dry. There are exceptions like the coastal climate of California or the

West Canaries, particularly the famous valley of Orotava of Tenerife.

The season that is considered most alien to the idea of a comfort climate in the temperate climates is of course winter. But winter may also have a favorable influence on man. By dressing appropriately in the open air, he surrounds his body with the effective coating of a warm climate. Moreover, the humidity is optimal in winter. In the humid tropics, the well-to-do can have single rooms or a whole house or a car air-conditioned, and thereby enjoy optimal temperature requirements, but in general it is more difficult to control the humidity of the air. In the open air, the inhabitant of the tropics cannot escape the unfavorable conditions by a change of clothing. For work, especially for mental work, the temperate zones therefore have always been the most suitable latitudes. This is reflected in the superior development of science, art, culture, techniques and civilization generally. The positive climatic influence specifically implies greater life expectancy. Thus, the peoples of the temperate latitudes have a much longer life span than the peoples of the tropics, whatever their race.

There are interesting parallels between human life and soil development. For soil development, the pause of winter in temperate zones is not disadvantageous. Moreover, it is not always a real pause. In areas where the snow falls on unfrozen soil, the soil life continues, particularly that of the majority of the soil animals. Even where the soil does freeze, the effect is rarely disadvantageous. Frost has a very important effect on the soil micromorphology, for it produces loose fabric types and aggregate formation. Frost in soils represents, on the one hand, a drop in temperature; but, on the other, it equals a form of drying. Changes in humidity are also the cause of changes in micromorphology, and consequently in the macroscopic soil structure. Frost and humidity are assumed to be the strongest agents in the phenomenon of "brown earthening" (*braune Vererdung*). The improvement in structure leads to an improvement in aeration, in water transmission and water capacity and in the biology of the soil. This is why the soils with the highest biological activity occur in the temperate zones—soils such as the Chernozems and the eutrophic Braunerde, with deep humus horizons, the best humus forms, high nutrient content, sponge-like structure, and a superior and complex soil life.

In looking for an example of a development sequence on silicate rock that could be applied as a standard sequence for comparative research, it was necessary to search among soils of the temperate zone, because here occur the longest development series and the most differentiated stages. Such soils are not found in the humid tropics, because there the initial stages are very short and indistinct. In the cold climates the initial stages are well developed, but the final stages are either non-existent or incompletely developed. Another requirement for our sequence was that it represent a strictly monogenetic development, one not influenced by climates of earlier periods or by other parent rocks.

The development sequence was chosen on the basis of studies conducted on granite in the Austrian hill section. The soils are typical of the Bohemian Forest of southern Bohemia, the Bavarian Forest and the northwest of Lower Austria.

SOIL DEVELOPMENT ON GRANITE

The soils of the granite area of Lower Austria are not the most fertile of the country. This is mainly because of the composition of the rock. On the southern border of the granite area a gneiss area begins. Both areas are inhabited by peasants. A folksaying of the countryside declares, "Where the granite begins, there ends intelligence." This is a simple notion because the reason for the difference in productivity is that two effective constituents are lacking in the granite but are present in the gneiss—biotite and the plagioclases.

The granites are acid and contain orthoclase, muscovite and a large amount of quartz. At best, a kind of Semipodzol (podzolic Braunerde) forms at the end of the mesotrophic Braunerde in the granite area. Crop yields in the gneiss area are much higher and it is therefore suitable for long-term farm expenditures and improvements that are often not feasible in the granite section. We chose the development sequence on granite because it is more interesting, although it leads to stages that display a reverse development, that is to the formation of soils that are lower in biological value and productivity than the preceding stages.

The granite development sequence also suited our purposes because granite is a rock of high distribution, found in abundance in many different regions of the earth. Furthermore, it displays marked changes as a result of different environmental conditions even before real soil development has begun. According to the region, therefore, characteristic geomorphological differences are apparent. In the humid tropics, granite is subject to a very deep and almost complete chemical weathering that gives rise to the formation of clayey or loamy masses that can be easily cut or dug with a spade, although their rock structure remains completely preserved. This affects the geomorphology, for erosion attacks not only the soils but also the rock surfaces; a flattened relief is produced as a result of the constant strong tendency to removal of the elevated parts of the landscape and to a filling up of the depressions.

In the summer-dry regions, a deep loosening of the rock structure takes place and, in consequence, is succeeded by the formation of granite pyramids, towers and citadels surrounded by wide, apron-like accumulations of rock débris. In the humid temperate zone, the granite exposed to air is transformed into a variety of rounded forms with hard, smooth surfaces, among which prevail those of huge woolbags and mattresses (Figure 6).

LIFE FORMS ON BARE GRANITE SURFACES

Life begins on granite as it does on other rocks, with the colonization of algae, bacteria, fungi, protozoa, and other lower forms on wet areas. Proliferating in wet habitats near streams and lakes is a bright-red alga that spreads like a lichen and has the odor of violets. It is *Trentepohlia iolithus,* the violet stone alga. But the first colonization of importance is produced by crustose lichens (epilithic genera). The principal lichens are *Rhizocarpon geographicum,* the map lichen, which displays bright-green map-like areas bordered by black seams. Then, there are several black species of *Lecanora.* These crustose lichens produce conditions favorable to the growth of other lichens which are readily able to collect organic and mineral detritus in a kind of humus layer: for example, the leaf lichen, *Peltigera* sp., and the reindeer lichen, *Cladonia rangiferina.* Soon mosses appear: first *Hedwigia ciliata,* which gives rise to the formation of a thin moder layer. The moss cushions permit the development of grasses, and these spread rapidly and have a good effect on humus formation. The humus layer grows considerably in thickness and consistency and develops into a moder rich in minerals (Plate 4C). The soil formed under these conditions is a Moder-Ranker, with the best development a mull-like Ranker (Plate 4E).

These developments are produced outside the forest, but sooner or later the open area reverts to forest, and in its final stage to a spruce forest *(Picea excelsa)*. Most of the bare rock surfaces are therefore situated under an old forest cover and undergo a very special development under more humid environmental conditions and the influence of reduced evaporation, reduced radiation and reduced wind action. The mosses under the shade of the forest may attain an extremely prolific growth. The main mosses are *Hylocomium splendens,* a storied branch moss; *Hylocomium triquetrum,* the garland moss; and *Hypnum crista castrensis,* the fir branch moss. Under the moss cover is a thin, blackish moder layer.

The growth of the mosses may become so luxuriant that it changes the nature of the soil development. Terrestric soil formation may continue, but the strong growth of sphagnum mosses can lead to a kind of peat moss formation. With constant excess of water, this develops into a semi-terrestric humus soil, sphagnum moss peat.

If the terrestric development progresses, the cushion cover is gradually colonized by shrubs of blueberries (*Vaccinium myrtillus*) and cowberries (*Vaccinium vitis idaea*). The shrub cover becomes very dense and has a marked influence on the humus formation. The humus horizon grows considerably and turns from a silicate moder to an acid mull-like moder—a very good seedbed for the propagation of spruce (acid Moder-Ranker to acid mull-like Moder-Ranker). The formation of a (B) horizon is a slow process. It gradually leads to the development of an acid Braunerde. The (B) horizon of this soil is completely different from that of Terra fusca. The clay content is small, the abundant iron hydroxide amorphous and flocculated, the fabric loose and rich in cavities. The A horizon contains an abundance of arthropod droppings, but their shape is unstable. With increasing humidity, they melt together and form blackish-brown, shapeless substances which are easily silted up and washed by the acid soil solution.

At this point the development goes downward, characterized by soil varieties with definite dystrophic humus forms. The humus horizons grow in thickness, but the value of the humus is diminished because it is less decomposed and less humified and its acidity has increased. Initial leaching of organic and inorganic decomposition products brings about the appearance of completely washed mineral grains in the humus horizon, and these begin to accumulate into disconnected patches or

Figure 6. Types of Granite Weathering

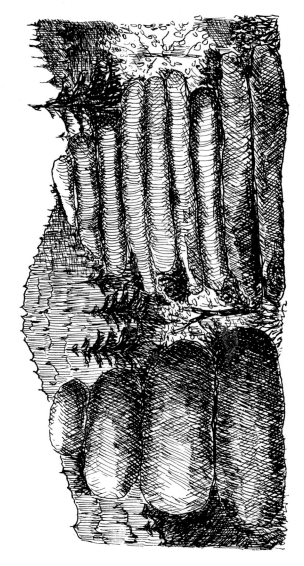

6A. Woolbag and mattress forms produced in humid temperate regions. Shown here is a spruce forest area (like the Bohemian Forest) with Semipodzol as the leading soil type. Moss and blueberry cushions with Proto-Ranker and mull-like Ranker develop on the bare rock surfaces.

Figure 6. (cont'd)

6B. Tower and citadel form with aprons of stone rubble formed in the mountain belt of the dry middle subzone of the Mediterranean. The dark areas are mats of the mountain broom *Sarothamnus purgans* with Tangel Ranker.

65

brims at the lower border of the humus layer. The profile now is that of a Semipodzol or podzolic Braunerde.

In dense old spruce forests and cold habitats, the blueberry and cowberry vegetation decreases. The effect of the spruce litter becomes more pronounced, the acidity increases and the humus form is transformed into raw humus (Plate 4A and 4B). The activity of the arthropods in the soil declines; fungi of deep-brown mycelium invade the plant residues and brown or redden the tissues without essential destruction of the cell structures of needles or roots (Plate 1C). The humified layer is thin. The leaching of brown humus sols and mineral colloids is very high, and this produces grayish to whitish bleached sand layers, followed by a coffee-brown B_h horizon (Iron Humus Podzol). The color of the soil solution is deep brown because of the large content of sols of fulvic acids. These humus sols invade the streams and ponds of the spruce forest. They come primarily from the terrestrial raw humus formation, but also from the highly dystrophic peat moss formations which may develop side by side with the podzols (Kubiëna, 1953a). Under the brown water of the streams and ponds, there is produced the subaqueous soil formation corresponding to the former, the brown, biologically very inactive Dy.

The full terrestrial development sequence is projected as follows:

(1) raw soil: very rare, almost nonexistent.

(2) Proto-Ranker: under *Peltigera* sp., or moss vegetation (*Hedwigia* sp., *Hylocomium splendens* and *triquetrum, Hypnum crista castrensis;* with grass vegetation: mull-like moder.)

(3) acid Moder-Ranker to mull-like Moder-Ranker: under *Vaccinium myrtillus* and *Vaccinium vitis idaea.*

(4) acid Braunerde: with growing age and increased chemical weathering.

(5) Semipodzol (podzolic Braunerde): influence of the spruce forest.

(6) Iron Humus Podzol: influence of raw humus and bleached sand formation. Heavily aged development stage. (Only in certain cold and biologically unfavorable habitats.)

X

Hypsometric Changes of Micromorphology on Silicate Rocks: Global and Local Influences

Changes of soil morphology and micromorphology in space, depending chiefly on the changes in life forms and in the environmental conditions of the soil habitat, are first of all determined by latitude and altitude. To these must be added the influences of continentality or oceanity (whether the habitat is centrally or peripherally situated on a continent) and of a western or eastern position. Such are the factors brought together in a system of geographic research by Lautensach (1952). But as soon as one endeavors to establish definite boundaries in drawing maps of zones, one perceives that soil development sometimes deviates considerably within the limits established. These deviations are due to local influences.

Spain is known for its unusual soil formations, and it is a land in which local influences are very important. In the last decade the late José María Albareda of Madrid was especially interested in that subject. I had the benefit of accompanying that dear colleague and friend on numerous excursions in the high mountains of Spain, in the Sierra de Guadarrama, the Somosierra, the Sierra de Gredos, the Serranía de Ronda, the Sierra de Cazorla, the Sierra Morena, the Sierra de Jaca, the Cantabrian Mountains and the Pyrenees. Albareda liked to take foreign guests to the Puerto de Navacerrada and, after a short walk on the slope of the Guadarramas, to show them the border of the Ranker zone on granite at an altitude of 1,900 m. (6,230 ft.).

In his last years he was deeply engaged in soil studies in the Pyrenees and intended to prepare a paper on that subject for the next Working Meeting on Soil Micromorphology, which was to take place in Madrid under his initiative and guidance. This paper would have been devoted to the hypsometric change of soil morphology and micromorphology in the high mountains of Spain. Further scientific excursions, in which more frame samples were to be taken, had been planned for the spring of 1966. Alas, Albareda died suddenly.

In order to recognize the distinctive features of a country and the significance of the local influences, one needs a general scheme that can be regarded as normal, that is, some standard for comparison. It had been agreed between us that I would try to trace such a general scheme, basing myself on the experiences of world-wide travel and research (Kubiëna, 1948, 1953a, 1967). This will be the subject of the following chapters. Their substance was also published in an Albareda memorial volume honoring this dear colleague and friend, his great professional and human qualities (Kubiëna, et al., 1967).

Among the local influences on soil development, the following can be regarded as the most significant: the parent material; age and maturity; human activity; relief (whether situated in depressions or on summits or slopes, steepness, south or north orientation of slopes, and so on); water conditions (proximity of springs, streams, lakes, and so on); and local high mountain climatic deviations (whether situated in a dry, semidry or humid high mountain zone).

In the present chapter the first three influences will be discussed.

THE PARENT MATERIAL

The parent material is of enormous importance in determining the zonality and the kind of soil formation. In fact, maps showing general rules must be limited to a specific parent rock or group of parent rocks. The greatest differences hinge on whether calcareous rocks or silicate rocks are present. Unfortunately, high limestone mountains have a much smaller distribution than mountains of silicate rocks. Comparative research on the hypsometric change in soil morphology on limestone in different latitudes is therefore more difficult to perform. I had the advantage of great familiarity with the soils of the high limestone mountains of the Austrian Alps, which are well developed. Thus the catena of the zonality of limestone soils affords a good insight into the great differences in soil development on limestone as compared with that on silicate rocks.

In excursions in the Spanish limestone mountains Albareda and I found very typical differences in the development sequences compared with those in the Austrian Alps, differences which offer a promising subject for future investigations. Also, the influence of different parent material within the group of silicate rocks is very marked in Spain. Thus, on the graphite schists of the summit of the Picacho de Veleta of the Sierra Nevada, whose content of silicate minerals is small, the border of the raw soils can be reached at an altitude of 3,600 m. (11,800 ft.); this is not the case with chlorite schist, which develops alpine Sod Braunerde at the same altitude. The summit of the Peñalara of the Sierra de Guadarrama is composed of granite without biotites, and the soils are alpine Rankers throughout.

THE INFLUENCE OF AGE AND MATURITY

Soil development is produced in a sequence of different stages, starting with the simplest and continuing gradually to more organized and complicated units. These units are represented by different soil types or subtypes. By this progression, there may be present in a given zone of soil development not only the final type of the development sequence but also the soil types of the transition or even the initial stages. One important characterizing feature of a given zone is whether only the initial and/or intermediate stages of a soil are present, or whether the climax formation is also present.

In the case of relict soils their morphology may be largely determined by the age factor quite independently of the environmental conditions of the present. To pedologists it is more and more apparent that paleosoils exist in the alpine zone, such soils having developed under tropical or subtropical conditions before the vaulting of the landscape to higher altitudes. These soils now display characteristics that are completely different from those of the recent soils formed under high mountain conditions. Thus, some soils possess certain characteristics that they developed in the Tertiary age. Geological history is also responsible for the morphology and micromorphology of certain soils. Lautensach (1952) was aware of the influence of age or time on soil development, but by the concept of "change of forms" (*Formenwandel*) he purposely wanted to limit himself to the change in space, without considering the change in time. The latter he regards as the province of the historical sciences. Geography, by his concept, is the consideration of the dynamic present, although the geographer may have "to consider genetic influences of the past upon a given geographical substance before starting to investigate the influence of the present."

Some characteristic transformation is always wrought on the paleosoils by the mountain climate; however, the paleopedologist knows that the morphological characteristics of the present may be minor compared with the inherited characteristics. Some former tropical or subtropical soils now exist in the form of Terra fusca varieties in the raised Tertiary landscapes of the Austrian Alps; in the form of Rotlehm varieties of the Andes of Peru (collected by H. Ellenberg); and in the form of Roterde varieties of the Sierra Nevada de Santa Marta of Colombia (Uhlig, 1966).

INFLUENCE OF HUMAN ACTIVITY

The presence or absence of certain soil types or the modes of their transformation may depend to a great extent on human activity and only in a degree on present environmental conditions. Spain offers a number of distinctive examples. Tertiary relict soils of Braunlehm or Rotlehm character are curiously well preserved on Spanish mountain slopes, whereas they do not exist in areas with less inclination. H. Hernández Pacheco demonstrated that these latter areas were precisely those that had been devoted to intensive, soil-destroying grain farming during the Roman colonization. From field investigations in the granite landscape of the Pedroches of the Sierra Morena it became clear to me that almost all the strongly weathered Tertiary relict soils had been carried away by soil erosion and replaced by the recent xeromorphic mediterranean Braunerde varieties. The Braunlehm and Rotlehm relicts on the steep slopes were preserved because these slopes had never been used as arable land and remained protected against soil erosion by the dense shrub cover of *Cistus* heaths (Kubiëna, 1955a).

The Braunerde varieties of central Spain are extremely dry and loose, and low in clay and humus content; they contain little soil life. These conditions are not directly caused by climatic conditions; rather they are attributable to human intervention, mainly the very tolerant regulations for sheep grazing of the Middle Ages and the continuous cultivation of grain crops in a dry climate. The mediterranean Braunerde has good mull formation, good soil life, well-developed humus horizons and well-preserved soil profiles in areas which formerly had *Quercus ilex* forests or areas which still have some groves of *Quercus ballota* in which hogs are allowed to feed on the acorns and the grass cover.

The extraordinarily deep border of the Ranker belt on granite in the western section of the Sierra de Guadarrama, beginning at 1,900 m.

(6,230 ft.) above the Puerto de Navacerrada, may also be attributable to former human use of land in these regions and subsequent removal of the original soils by erosion.

This explains briefly the complexities of relict soil morphology of some selected soils which formed under normal drainage conditions in past geological ages. The subject of soil development obviously becomes even more complicated if one considers that soils are not formed just on well-drained land surfaces (terrestric soils); they develop also in areas where water has a tendency to stagnate inside the soil profile, or in areas which undergo temporary flooding during a part of the year (semiterrestric soils), or even in areas constantly under water (subaqueous soils).

The different soil types of these three different realms of soil development possess definite coordination properties; since there are definite interrelations between them they can tolerate temporary change of environments. It is also true, however, that subaqueous soils may change into semiterrestric soils and continue their development in the altered habitat, and finally finish as terrestric soils. Or vice versa. For a complete analysis of a landscape in regard to soil formation, the soils of all three realms of development must be considered; however, this is still a difficult task, and rules of soil development within the three realms must first be studied independently.

XI

Hypsometric Changes of Micromorphology on Silicate Rocks: The Pastertze Glacier Sector

Before attempting to describe complete catenas of soil development in high mountains, it may be useful to report on some soil development studies begun by Friedel (1934) and continued by the author in the Austrian Alps in the vicinity of the greatest Austrian ice field, the Pastertze glacier. This glacier is situated in the High Tauern of the Eastern Alps of southern Austria. Above it rises the highest Austrian mountain peak, the Grossglockner, 3,800 m. (12,460 ft.).

In the vicinity of a large glacier which spreads over large areas and whose movements are clearly marked by numerous well-dated moraines, the succession of the soil formations between the moraines is easily recognized and every stage of soil development can be dated accurately.

Figure 7 shows the location of the several moraines of the Pastertze glacier. The moraine of 1856 marks the highest advance of the glacier during the last century; the later moraines mark a successive retreat that has continued until very recent years, so that the glacier now stands far behind the moraine of 1934.

The Fernau moraine corresponds to the high stand of the glacier in the Middle Ages, some 360 years ago; the Eggessen moraine of the post-glacial period was deposited about 10,000 years ago. The more easterly Daun moraine, deposited about 12,000 years ago, lies outside the area of the sketch.

Here are some details on the rock formations of the area. The elevation known as Elisabeth Rock, 2,100 m. (6,890 ft.), which was at the time of the investigations (1939) completely free of ice, is composed of calcareous mica schist; so also are the depression Margaritze Glen, 2,020 m. (6,625 ft.), the elevated terrain known as Marx Meadow and another high area in the northeast corner. This rock consists of calcite, quartz, muscovite and some small grains of iron ores, mainly pyrite. The rock walls in the north-northwest corner above Möll Brook and the rock walls in the southeast corner south of Elisabeth Rock are composed of prasinite. This dense, bright-olive-green rock of striking appearance forms the near

Figure 7. Moraines of Different Ages and the Corresponding Soil Formations in the Vicinity of the Pastertze Glacier, Austria (after V. Paschinger)

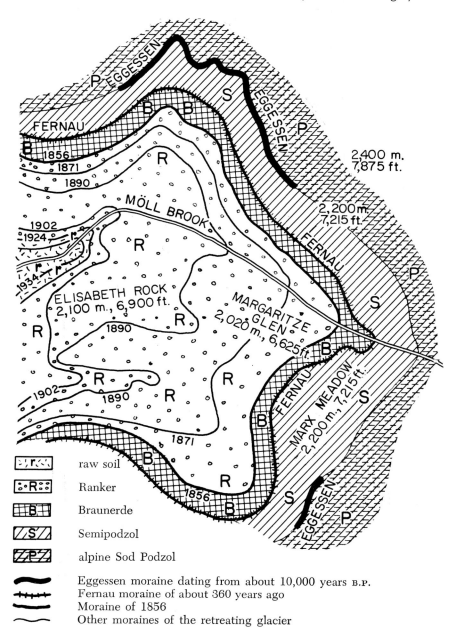

raw soil

Ranker

Braunerde

Semipodzol

alpine Sod Podzol

Eggessen moraine dating from about 10,000 years B.P.
Fernau moraine of about 360 years ago
Moraine of 1856
Other moraines of the retreating glacier

73

peak of the Grossglockner. It is composed of epidote and chlorite or epidote and hornblende. These rocks, or a mixture, are the principal parent materials of the moraines.

The soil development is more or less typical for soil on silicate rocks in the alpine grassland belt (Figure 8). Rock débris of calcareous mica schist behaves like débris of silicate rocks in that the calcite content is soon removed by solution. Beginning soil formation could be observed on the east slope of Elisabeth Rock. During the Middle Ages the rock was covered by a dense grassland, but when this cover was completely removed by the movement of the ice front during the late Middle Ages, it exposed a polished base rock whose surface had been variously scratched by the rock fragments frozen into the moving ice. When I was a young man, I had to climb up from Elisabeth Rock to reach the glacier front, but today I go down. In fact, the border of the glacier has retreated north-westward to a considerable distance from Elisabeth Rock. In front of the glacier front, at a time when it was above Elisabeth Rock, we observed a series of small moraines that had been laid down year by year (winter moraines). Since they were colonized only in patches by pioneer vegetation, the soil had no more than raw soil character.

These small annual moraines were preceded by the well-developed moraines of the retreating glacier of 1934, 1924 and 1902. Beyond Elisabeth Rock, laid down as a result of a previous much stronger retreat of the ice front, are the moraines of 1890 and 1871.

The most favored spots for plant colonization between the moraines were the small depressions that contained finer débris and had better water conditions. Here, although the formation of humus horizons takes place only in the patches of plant growth, a humus layer may be built up within four or five years. The humus form is a primitive moder that gives rise to a remarkable activity of small arthropods, mainly oribatids and collembola. The soil has the character of a Proto-Ranker. Beyond the moraine of 1856, we found everywhere a dense and completely closed grassland upon the undisturbed soil cover. The soil profile here had a marked (B) horizon, the profile of a Braunerde.

The moraine itself as a rule had no Braunerde; however, Braunerde has developed in certain spots, primarily where heavy sods of the former grassland had been removed by the ice and pushed onto the crest of the moraine dam. The dam itself, in altering moisture conditions, favored the development of a closed grass cover all over its surface. Buried sods of the former grassland were also found in the interior of the moraine.

Figure 8. Stages of Soil Development in the Vicinity of the Pastertze Glacier, Austria

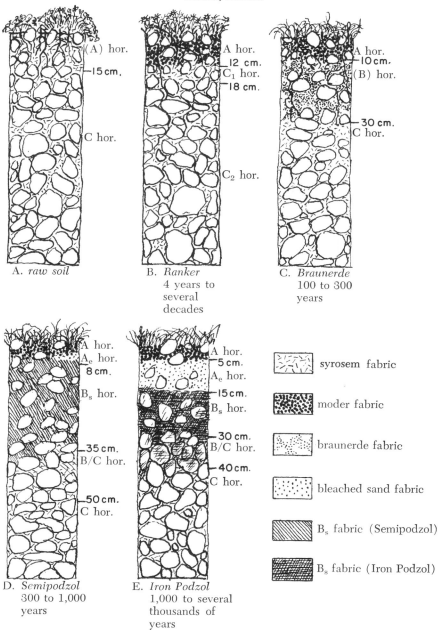

A. *raw soil*

B. *Ranker*
4 years to
several
decades

C. *Braunerde*
100 to 300
years

D. *Semipodzol*
300 to 1,000
years

E. *Iron Podzol*
1,000 to several
thousands of
years

syrosem fabric

moder fabric

braunerde fabric

bleached sand fabric

B$_s$ fabric (Semipodzol)

B$_s$ fabric (Iron Podzol)

The areas between the moraine of 1856 and the Fernau moraine contained Braunerde but no soil formation of more advanced development. This indicates that the development of Braunerde continued without change for about three hundred years. The time required for the development of a Braunerde profile is at least a hundred years. Beyond the Fernau moraine, podzolization became evident. The humus formation was an acid coarse moder. Leaching was clearly visible, as was also the formation of an illuvial B horizon. In spite of the more or less flocculated iron hydroxide still present, the structure was fairly dense. Near the remains of the Eggessen moraine, some formation of bleached sand was apparent, particularly where the dominant parent material was calcareous mica schist and its débris.

The real Podzol zone began beyond the line of the Eggessen moraine. The average time for the formation of alpine Podzols with well-developed bleached sand layers varies from a thousand to several thousand years. They are like those found everywhere in the grassland belt of the Austrian Alps—Iron Podzols with A, A_e, B_s, horizons (Figure 8).

The micromorphology of the soil types of the whole development sequence corresponds to those to be described in the following chapter. The shrub- and timberline in the vicinity of Möll Brook is lower than might be expected, evidently as a result of human activity.

XII

Hypsometric Changes of Micromorphology on Silicate Rocks: Development Catena of the Austrian Alps

Since there are great differences among noncalcareous rocks, it is very difficult to establish any general rules with respect to the development of soil types and their distribution in space. At one extreme are the siliceous rocks (siliceous schists, graphite schists, quartzites); at the other are basic silicate rocks (basalts, diabases, norites and basic silicate schists); in the middle are granites and acid gneisses of equivalent composition. The last, conveniently for our purposes, not only have a wide distribution but are common in different parts of the earth. It is with this particular group of silicate rocks that we shall be concerned in this chapter, and specifically with the catena of soil development on these rocks in the Alps. This catena is the most suitable for study because it is the most complete and the best developed.

Soil development studies on granite and gneiss in the Alps were first carried out by Pallmann and Haffter (1933) in the Upper Engadine of eastern Switzerland. They were continued by the author in the Austrian Alps, particularly in the High Tauern (Kubiëna, 1938), where these belts are situated at somewhat lower altitudes but are very similar in nature and succession to those of the Upper Engadine. The altitudes of the belts are lower because of the lower mass elevation of the Austrian Alps. Relevant data on the hypsometric changes of soil profile development are presented in Table 4 and Figure 10C.

LOWLAND AREA AND HILL BELT

Up to about 800 m. (2,625 ft.), grain crop production alternates with deciduous and pine forests on most of the arable land. The soil type changes according to local influences; however, Central European Braunerde is present everywhere and can be regarded as the most representative soil for this belt, particularly with respect to arable land. The humus formation is generally mull; in some cases, primarily in beech and pine forests, it is mull-like moder.

Table 4   Hypsometric Change of Soil Profile Development on Silicate Rocks in the Austrian Alps

Climax Formation	*Phases of Profile Development*
8. alpine silicate Råmark	(A)C
7. alpine Ranker (up to 3,000 m. = 9,840 ft.)	(A)C,[a] AC
6. alpine Sod Braunerde (up to 2,900 m. = 9,500 ft.)	(A)C, AC,[b] A(B)C
5. alpine Sod Podzol (up to 2,800 m. = 9,200 ft.)	(A)C, AC,[b] A(B)C,[c] ABC,[d] AA$_e$BC
4. subalpine Iron Humus Podzol (up to 2,100 m. = 6,900 ft.)	(A)C, AC, ABC,[d] AA$_e$B$_s$C,[e] AA$_e$B$_h$B$_s$C
3. forest Iron Podzol (up to 1,800 m. = 5,900 ft.)	[(A)C], AC,[f] ABC,[d] AA$_e$B$_s$C
2. Semipodzol (up to 1,000 m. = 3,300 ft.)	[(A)C], AC,[f] A(B)C,[c] ABC
1. Braunerde (up to 800 m. = 2,625 ft.)	[(A)C], AC,[f] A(B)C

[a] silicate raw soil (not climax-forming)
[b] alpine Ranker
[c] Braunerde
[d] Semipodzol
[e] Iron Podzol
[f] Proto-Ranker and mull-like Ranker

MONTANE BELT

This belt, extending to about 1,000 m. (3,300 ft.), is characterized by dense deciduous and pine forests in which Semipodzols, and, on some north slopes, Iron Podzols are formed. The representative soil type is Semipodzol. The lower subalpine belt, extending to about 1,800 m. (5,900 ft.), is covered mainly with spruce forests, and the leading soil type is Iron Podzol.

UPPER SUBALPINE BELT

This belt reaches to about 2,100 m. (6,900 ft.), and is characterized by larch and cembra pine forests in the upper part of a shrub belt with a dense plant cover of mountain roses (*Rhododendron ferrugineum*), blueberries (*Vaccinium myrtillus*), and cowberries (*Vaccinium vitis idaea*).

The larch and cembra pine forests have well-developed Iron Podzols with deep soil profiles and thick horizon formation. In the plant cover beneath the trees, blueberry bushes and mosses predominate. Together with the copious litter of the larches and cembra pines, they foster the development of unusually thick raw humus layers with a luxuriant growth of dark-brown fungus mycelia and a small amount of animal droppings. The thin H horizons form a sharp border above the mineral soil underneath, beginning with a deep, bleached sand horizon which continues into an olive-brown B_s horizon.

In the shrub belt outside the forest, blueberry and cowberry bushes and mountain roses constitute most of the plant cover of a rhodoretum extrasilvaticum. The content of animal droppings in the humus layer increases, and the content of dark fungal hyphae diminishes. The H layer is considerably thicker. Below the bleached sand layer is a coffee-brown B_h layer stained by accumulated humus substances, and this is underlain by an ocher-brown B_s horizon.

There are other dwarf shrubs, such as *Loiseleuria procumbens*, which grows in the form of a horizontal trellis nestled close to the soil surface. Sometimes the *Loiseleuria* (which may grow even into the alpine belt) contains a large amount of wind- and frost-resistant lichens, such as *Cetraria nivalis, C. crispa* and *C. cucullata*, the worm lichen *Thamnolia vermicularis*, and *Alectoria* species. In these areas, podzolization is considerably checked by the unfavorable climatic conditions of the habitats.

A very typical change in vegetation and soil formation occurs in the alpine belt, which reaches to an altitude of about 3,000 m. (9,840 ft.). This change is caused by the climatic conditions.

ENVIRONMENTAL CONDITIONS OF THE UPPER BELTS

Above the timber- and shrubline, temperatures are very much lower than in the corresponding lowlands. In the nival zone, it snows more frequently and more abundantly, and the snow remains on the ground for longer periods. The soil is frozen under the snow cover; only in summer does it thaw somewhat. The difference between winter and summer is nonetheless tremendous, even though summer snows are sometimes heavy.

The summer is very marked in the alpine grassland belt of the Austrian Alps. Hot sun, cool shade, high precipitation, frequent snowfall, long-lasting snow cover, thawing over frozen soil (in its frozen state the soil is impermeable)—all these conditions on silicate rock bring about the development of very characteristic soil varieties, with a distinct micromorphology.

ALPINE BELT

The dense grassland in the alpine belt is composed of a plant community in which the crooked sedge *Carex curvula*, with curved stems and leaves, prevails. Below this vegetation develop shallow Iron Podzols with thin and irregularly formed horizons, the alpine Sod Podzols. They are somewhat similar to the Nanopodzols, the dwarf Iron Podzols of subarctic forest tundra; however, they are typical grassland soils in the alpine belt. Each horizon is only a few centimeters thick.

SUBNIVAL BELT

At 2,800 m. (9,200 ft.), the alpine belt grades into the subnival belt, which extends up to about 3,000 m. (9,840 ft.). Here the closed grassland vegetation gradually opens and plant life is confined to cushion growth between areas of more or less bare rock débris. Some grassland occurs in a transition zone in which develop soils with A(B)C profiles, the alpine Sod Braunerde in which podzolization has completely stopped.

The Braunerde belt grades into a Ranker belt which forms a blackish humus horizon of mull-like silicate moder. This alpine Ranker represents the climax formation in this region. The climax Ranker belt is considerably more developed in the Central Alps of Switzerland and exhibits a marked development in the Mediterranean high mountains. In addition to low annual temperatures, the degree of continentality, the lower humidity and the higher mass elevation are of essential influence in producing this phenomenon.

NIVAL BELT

Beginning at about 3,000 m. (9,840 ft.), mechanical decay produced by rapid changes of temperature intensifies, whereas chemical weathering is greatly reduced. The vegetation is limited to a few lichens and weather-resistant plants. Snow and ice, bare rock and deposits of bare rock débris are characteristic of the landscape. In regard to soil development, the formation of humus horizons is very rare. The representative soil type is the alpine silicate raw soil or alpine silicate Råmark. The development sequence consists of only one member—the Råmark—which is both the initial and the final stage.

XIII

Hypsometric Changes of Micromorphology on Silicate Rocks: Mountain Catenas in Different Latitudes

Changes in the morphology and micromorphology of mountain soils are produced not only by the altitude but also by other influences, especially latitude. In lowland areas, beginning at the Equator and proceeding north to the polar region, we see, with some exceptions, a shortening of the development sequences. This shortening occurs in much the same way as the mountain sequences are shortened by increasing altitude.

The effect of changes in latitude is suggested in the idealized scheme in Figure 9, in which the planetarian change of morphology is indicated by gradual change of the profile formation. The circles in that scheme do not of course represent real parallels; the distances between them do not indicate their distance in nature, and the area of the inner circle does not correspond to the area of distribution of the arctic raw soils or terrestric soils in general. The scheme shows only the progression of the typical soils and the genetic sequences that are representative for a global scheme.

This scheme provides the basis for constructing a mountain catena within any given latitude. The soil catena of a mountain must fit into one of its circles since it must continue the genetic sequence of the lowland sequence of a given latitude.

Proceeding from the northern temperate zone towards the pole, all sequences of higher altitude are bound to be shorter than that of the foothill region. Since from here on, with increasing latitude, the genetic sequences become abridged, so also the mountain catenas are shortened, until finally the alpine Råmark and the arctic Råmark grade into each other and become identical. The whole genetic sequence, initial and final stage, consists of only one member. With increasing altitude there can be only Råmark, ice, snow, bare rock or rock débris.

The two thick lines delineating the mediterranean zone in Figure 9 separate two sections of the hemisphere. In the northern section, the arctic and temperate zones, by their close interrelations, form a clear unit. In the southern section, the tropical and subtropical zones follow different rules that will be discussed further on. In the mediterranean zone, the characteristics of the sections to the north and south merge and form a kind of intermediate zone of soil formation. This zone also represents the border in which the great climatic movements of the past, clearly recorded by the remains of paleosoils, came to a standstill—the advances and retreats of the polar climate during the Pleistocene, to the south and back, and the advances and retreats of the Rotlehm border from the tropics to the north and vice-versa.

But in every latitude, all mountain sequences are shorter than the genetic sequence of the lowland area. The long sequences of the lowlands generally are characterized by a kind of retrogressive development in which the final stages, produced by aging, represent formations that are biologically less valuable than the preceding stages of development. In the middle latitudes of the temperate zone, which are favored by optimal environmental conditions, this aging does not occur in the lowland area or in the hill belt; it begins in the lower subalpine belt. The shortening of the catenas as a result of increasing latitude takes place in much the same way as the shortening as a result of altitude that starts in the northern temperate zone.

The complete catena of the hypsometric change of profile morphology on acid silicate rocks of the Austrian Alps was presented in Table 4 (Chapter XII). Here, the horizon designations of the profiles indicate the complete genetic sequences; also given are the climax formations, as in Figure 10C.

In this catena, the sequence of the subalpine Iron Humus Podzol indicates a development that is produced by the transformation of a forest Iron Podzol at the upper border of the larch and cembra pine forest.

The mountain schemes in Figure 10 are in essence built up from the facts presented in Figure 9 showing the principal genetic sequences of the lowland areas of the Northern Hemisphere. The mountain schemes apply to different latitudes, beginning at the Equator and ending with the high Arctic, and they show the change of soil types with increasing altitude. The contrast between the climates of the northern and southern sections (both divided by the mediterranean zone) is reflected in the composition of their corresponding high mountain belts. Environmental conditions in the mountains of the northern and the southern sections are indeed entirely different. In the north, a cold winter alternates with a well-marked summer. This seasonal change is minimal in the high mountains of the tropics.

Figure 9. Planetarian Change of Soil Development on Silicate Rocks in Various Areas of the Northern Hemisphere, in Centropolar Presentation

and

Figure 10. Hypsometric Change of Soil Development on Silicate Rocks in Various Areas of the Northern Hemisphere at Different Latitudes (idealized scheme). Letters correspond to the circled letters in Figure 9, indicating the planetarian zones in which the mountains of Figure 10 are situated.

Figure 9. Planetarian Change of Soil Development on Silicate Rocks in Various
Areas of the Northern Hemisphere, in Centropolar Presentation

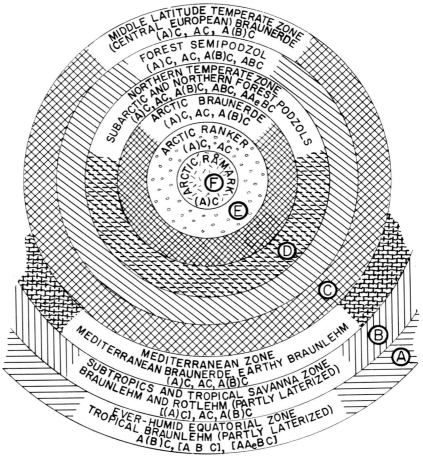

EQUATOR

Figure 10. Hypsometric Change of Soil Development on Silicate Rocks in Various Areas of the Northern Hemisphere at Different Latitudes (idealized scheme). Letters correspond to the circled letters in Figure 9, indicating the planetarian zones in which the mountains of Figure 10 are situated.

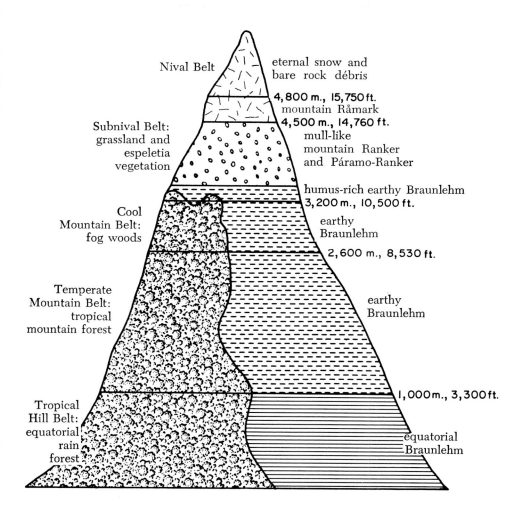

Figure 10A. Colombian Andes, ca. 5° N

Figure 10 (cont'd)

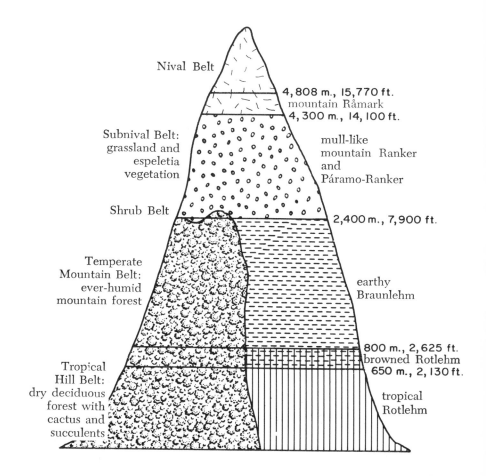

Nival Belt

4,808 m., 15,770 ft.
mountain Råmark
4,300 m., 14,100 ft.

Subnival Belt:
grassland and
espeletia
vegetation

mull-like
mountain Ranker
and
Páramo-Ranker

Shrub Belt

2,400 m., 7,900 ft.

Temperate
Mountain Belt:
ever-humid
mountain forest

earthy
Braunlehm

800 m., 2,625 ft.
browned Rotlehm
650 m., 2,130 ft.

Tropical
Hill Belt:
dry deciduous
forest with
cactus and
succulents

tropical
Rotlehm

Figure 10B. Colombian Andes, ca. 11° N

86

Figure 10 (cont'd)

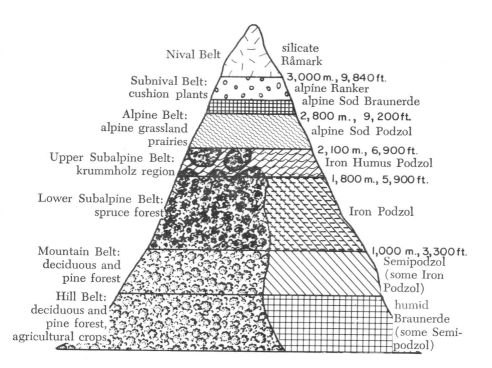

Figure 10C. Austrian Alps, ca. 47° N

Figure 10 (cont'd)

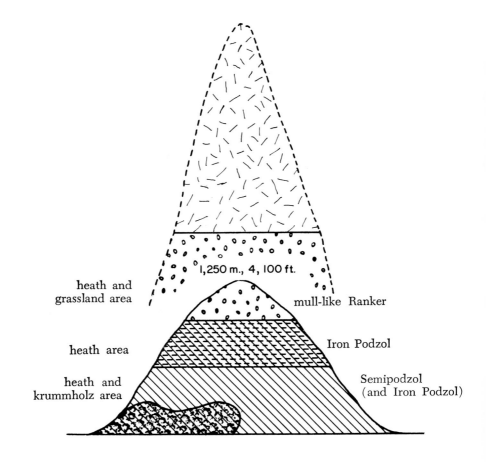

Figure 10D. Cairngorm, Scotland, ca. 57° N

Figure 10 (cont'd)

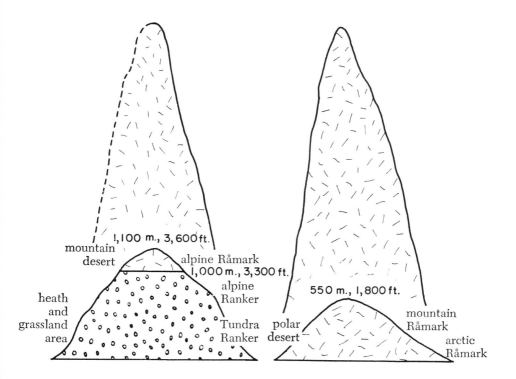

Figure 10E. Tundra of
Northern Norway, ca. 70° N

Figure 10F. Peary Land,
Northern Greenland, ca. 83° N

The abridgment of the genetic sequences with increasing latitude in the northern section has important economic implications. The landscapes of the lower soil belts in the Alps gradually disappear if one continues towards the north. The borders of the nival, subnival and alpine belts are found at lower altitudes. Thus, on the granite mountain of the Cairngorm in Scotland, one sees some forest cover in the lowland area, but only shrubs in the lowest mountain belt, and a Semipodzol in the hill section (Figure 10D). With increasing altitude, a belt of Podzol formation can be seen; it is followed by a belt of mull-like Ranker with grassland vegetation above the 1,000 m. (3,300 ft.) line. Since the Cairngorm is only 1,250 m. (4,100 ft.) high, no additional climax soil formation is developed. If it were higher, the border of a Råmark belt presumably would be reached at about 1,800 m. (5,900 ft.). If the Cairngorm attained a height equal to the summits of the High Tauern in Austria, the Råmark would be present uniformly to the mountain top.

However, the border of the Råmark belt is easily reached farther north in mountains of the same height as the Cairngorm; this is indicated by a mountain scheme of the tundra zone of northern Norway in Figure 10E. It consists of a Ranker belt; Braunerde, Semipodzols or Podzols do not occur. Råmark formation begins at 1,100 m. (3,600 ft.), and continues to the top.

In Peary Land, at about 83° north latitude, the hills have only arctic raw soil (according to an oral communication from B. Fristrup) (Figure 10F).

It must be emphasized again that the series of mountain schemes of Figure 10 cannot present all possible changes of alpine soil formation with increasing latitude and altitude, and that such is not intended by the figure. Its only purpose is to present a series of schemes sufficiently typical to be useful as a common basis for comparative research. In this sense, it attempts to express only the combined effect of latitude and altitude upon the kind and intensity of additional influences (as indicated in Chapter X). It can be said that in general the kind of soil formations and the succession of the soil belts follow the rules presented. In essence, the differences are determined by the altitudes of the belts, and this can only be recognized if one has a standard for comparison.

Additional influences, particularly increasing humidity, greatly determine the kind of semiterrestric soils that will occur in addition to the above-mentioned terrestrial soils—whether in the form of Pseudogley, Gley, Anmoor or peat formation. Since this chapter is confined to terrestrial soil development, semiterrestric soils must be left for later investigations.

It may be noted that semiterrestric soils are much more developed in the Arctic at the present time than they were in the Pleistocene period, when the polar regions were much drier. The differences between the northern and the southern zones indicated in Figure 10 are evidenced by the occurrence of entirely different soil types. The equatorial climate of the lowland area brings about a soil formation that has undergone perhaps the most intensive chemical weathering on earth; it contains a large amount of amorphous mineral decomposition products. A very dense matrix contains amorphous silica, highly decomposed clay substances and peptized amorphous iron hydroxide. The soil type, very typical for that area, is tropical Braunlehm. The Braunlehm soil continues even into the cooler mountain belts of the Colombian Andes (Figure 10B), although in the form of earthy varieties. This results from the lack of a cold season and hence no seasonal snow cover.

In the Alps, the upper border of the forest grades into a krummholz belt (Figure 10C) which contains trees deformed by weather factors. Among these trees there is a luxuriant brush vegetation which exerts a very special influence on humus and soil formation. The krummholz area exists because the long-lasting winter snow cover protects the dwarfed and deformed trees against frost, wind action and drying.

No krummholz is found in the tropical high mountains. The mountain forest, very rich in epiphytes, mosses and lichens on the trees and in shrubs underneath them, grades out directly into the grassland and espeletia belt. The humidity is very high; heavy fog formations and long-lasting drip rains are common. The lack of winter causes the deep humus soils of the Páramo-Ranker in the subnival belt to contain elements of earthy Braunlehm, or even completely unflocculated braunlehm fabric as fillings in cracks or former root channels. The Páramo-Ranker has an active arthropod population, as does the Pitch-Rendzina of the limestone Alps. The humus form is a mull-like moder which attains a thickness of 88 to 127 cm. (35 to 50 inches).

The Páramos north of the Equator are more humid than those south of it, which change gradually to the periodically dry puna regions further south. However, even the dryer Páramo of Ecuador has a considerable amount of Braunlehm elements with the micromorphology of the Páramo-Ranker, as has been shown by the investigations of Frei (1964). The blackish peptized fillings of conducting channels that he describes are also known in prairie soils rich in Braunlehm elements. The micromorphology of his Brunizem varieties has the character of an earthy Braunlehm (earthy Braunlehm-Lessivé).

Soil formation in the tropics differs from that in the Alps partly because of the much lower situation of the soil belts in Europe. Whereas the raw soil belt in the Austrian Alps is reached at an altitude of 3,000 m. (9,840 ft.), in the equatorial Andes it begins at about 4,500 m. (14,760 ft.). The timberline in the Austrian Alps is situated at 1,800 m. (5,900 ft.), in the equatorial Andes at 3,200 m. (10,500 ft.) or higher. Whereas podzolization is a very typical process of the subalpine and alpine belts, slight podzolization may occur only occasionally in the tropical high mountains, although leaching may be quite common in sandy Braunlehm profiles of the foothill region. A notable characteristic is the width of the soil belts. The Ranker belt in the Austrian Alps is only 150 to 200 m. (500 to 650 ft.) wide; the Ranker belt of the equatorial Andes is 1,000 to 1,500 m. (3,300 to 5,000 ft.) wide.

Farther north, where the lowland zone is characterized by alternating dry and wet seasons and by the formation of rubified soils (Rotlehm, Roterde), the change of soil morphology is somewhat altered, as indicated for the Colombian Andes in Figure 10B. The Rotlehm of the lowland region with dry deciduous forest reaches to about 650 m. (2,130 ft.) and is followed by a transition belt of browned Rotlehm (verbraunter Rotlehm) with humid deciduous forest. The wide belt of ever-humid mountain forest has earthy Braunlehm. With the lower tree line at 2,400 m. (7,870 ft.), the grassland belt with Páramo-Ranker and mull-like Ranker begins considerably lower than in the lower latitudes of the Colombian Andes (Figure 10A). The width of the soil belt is also a characteristic feature. Th⁺ is why the complete set of soil belts can be observed only in the highest mountains. In many tropical mountains, the highest elevations are only a little higher than the shrub belt.

XIV

Soils of the Foothill Belt and the Lowland Areas of the Temperate Zone

In the temperate zone, as well as in the tropics, a sequence of soil varieties at different altitudes can be demonstrated by tridimensional presentation. Knowledge of soil development in the lowland areas is an important basis for a study of this sequence.

The great diversity of the climates of the temperate zone is mainly attributable to the great differences in temperature during the winter and to the differences in humidity. In spite of these differences, terrestric soil formation is very similar all over the zone, that is, from the Mediterranean to the Arctic zone. All the temperate climates have something in common that is decisive for soil development: the mild summer. The hard winter characteristic of some parts of the temperate zone has in general no disadvantageous influence on soil formation, but rather is favorable for the development of desirable structures and micromorphological characteristics.

Winter temperatures and humidity are determined primarily by the degree of continentality. With reference to the importance of these factors for soil development, and without considering local influences, the temperate zone can be divided into the following subzones: the humid temperate subzone (England, Austria); the humid winter-cold subzone (northern Eurasia); the dry winter-cold subzone (East Siberia).

(1) The humid temperate subzone comprises two sections—the oceanic and the semicontinental. The oceanic section, of which Western Europe is an example, has more cloud formation than the semicontinental section and high precipitation (the highest is in the Lake Region of England). Typical for this section are western France, the Netherlands, Belgium and northern Germany.*

Typical countries for the semicontinental section are Austria, western Hungary, northern Italy, Switzerland, Yugoslavia, Czechoslovakia, southern Germany and large parts of Poland. The differences between summer

---

*Some people call the summer in Hamburg a green winter. The differences between summer and winter are less pronounced than in the semicontinental sector.

and winter are greater in this section; cloud formation is lowest in summer. The annual precipitation is 500 to 800 mm. (20 to 32 inches).

To the humid temperate subzone belong also the eastern states and provinces of North America, Uruguay, central Argentina, southern Japan, the southern portion of Korea, southeast Australia, Tasmania, and South Island of New Zealand.

(2) In the humid winter-cold subzone the contrast between winter and summer is very great. The winter is long and cold. The annual precipitation is 500 to 700 mm. (20 to 28 inches). With increasing continentality, the temperature difference between winter and summer increases. Winter temperatures in West Siberia range from $-36°$ to $-40°C.$ ($-33°$ to $-40°F.$).

Typical for this subzone are Norway, Sweden (with exception of the southern tip of the peninsula), Finland, eastern Poland, northern Russia, West Siberia, the Great Lakes, and the Northeast of the United States, southern and middle Canada, the coastal region of East Siberia and northern Japan. No such subzone exists in the Southern Hemisphere.

(3) The dry winter-cold subzone can be designated as a temperate region only because of the warm summers. The winters are extremely cold (the lowest temperature in my own experience is $-56°C.$ ($-68°F.$) in East Siberia). The low winter temperatures are a result of the great continentality. The annual precipitation is very low. In the dry Kirghiz Steppe it amounts to only 120 mm. (4.8 inches). In most of East Siberia the soil is frozen to great depths. Still, for life and soil development, it is not the winter but the warm summer that is decisive. Soil and vegetation develop in much the same way as in the beginning stages in the semicontinental temperate zone of Europe. Typical for the dry winter-cold subzone are East Siberia, Mongolia, Manchuria and North China.

I myself worked four years in East Siberia and three months in central Manchuria. The temperatures in the sunshine and in the shade differ greatly; so also do the temperatures of day and night. Winds are rare and weak; the air is dry; the sky in both summer and winter is clear and almost always without clouds; and the light radiation is intense.

This kind of climate is found only in the above countries of the world. This is because such a high degree of continentality is reached nowhere except in Central East Asia. Similar conditions are found in some parts of the United States, particularly in parts of Montana, but they do not approach those of the East Siberian winter. However, in traveling through Montana in 1933, I saw landscapes and soil formation very much like those of some parts of East Siberia.

DEGREE OF CONTINENTALITY

The great contrasts in the climates of the temperate zone, and the need to subdivide it, show that it is not enough to consider merely latitude and altitude, strong as their influences may be, to understand the great diversity of soil formation. A third main controlling influence in the temperate zone is continentality. In the humid temperate subzone a marked difference in soil formation can be observed as one moves nearer the ocean or towards the center of the continent. Greater continentality means not only a decrease in humidity but also a considerable increase in coldness during the winter.

SOIL FORMATION OF THE HUMID TEMPERATE ZONE

If Braunlehm is the most representative soil type of the humid tropics, humid Braunerde or Ramann-Braunerde may be regarded as the most representative soil type of the humid temperate subzone. Ramann-Braunerde (Ramann, 1905) has sometimes been designated in the literature as Central European Braunerde. This indicates that the best varieties occur, not in the oceanic section, but in the semicontinental section, where the difference between summer and winter (which is colder, steadier, and longer) is considerably greater. Furthermore the pathways of transformation and its processes are different for the paleosoils in the different sections of the humid temperate subzone.

In contrast with tropical Braunlehm, Braunerde has a very low capability for transformation. In its natural environments, Braunerde is a very stable climax formation. This is particularly true of the eutrophic and mesotrophic varieties even if dystrophic humus formation is produced upon them. Podzolization occurs only in some oligotrophic varieties which, so to speak, already contain the predisposition for podzolization. In spite of that fact, many varieties of braunerde-like forms occur in the temperate zones even though typical Braunerde is rare (it occurs mostly in localities where there is fresh parent material which does not contain remnants of former soil formation).

Great variety results because of the presence of a multiplicity of transformation forms from relict soils and because of the occurrence of mixed pre-stages of some polygenetic soil development, such as Parabraunerde,

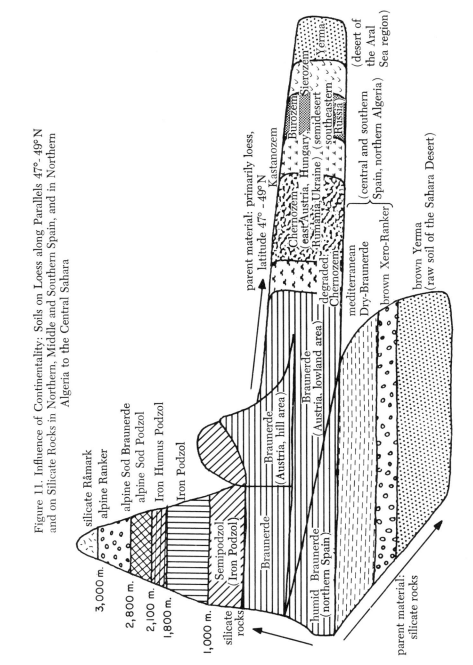

Figure 11. Influence of Continentality: Soils on Loess along Parallels 47°- 49° N and on Silicate Rocks in Northern, Middle and Southern Spain, and in Northern Algeria to the Central Sahara

Braunerde-Lessivés, brown Prairie soils, earthy Braunlehm, earthy Terra fusca, browned Rotlehm, browned Terra rossa, browned Roterde and others.

INFLUENCE OF CONTINENTALITY

How much continentality influences soil development is demonstrated in Figure 11, which gives an idealized lowland zonation beginning with the Braunerde area in the foothill belt and the surrounding area of the Eastern Alps. Next to a high mountain of acid silicate rock is a wooded spur hill of about 1,015 m. (3,330 ft.) (parent material: sandstone). The Braunerde belt reaches higher up on the east and south slope than it does on the west and north slope, as does the Semipodzol and the Podzol belt. Podzols occur generally on the north slope, at an altitude of about 1,000 m.

If we proceed eastward along the parallels of 47° to 49° N, that is, towards the center of the Eurasian continent, we find a different soil development. Although we move along in about the same latitude and stay at about the same altitude, we find not only different soils but also another shortening of the development sequences until finally we reach a sequence with only one member, the raw soil of the dry desert, or Yerma. Thus, continentality produces a phenomenon similar to that produced by altitude—by climbing a high mountain through the different mountain belts—or by latitude—by moving towards the polar region through the different planetarian zones. The longest sequence with the largest number of development stages is the Braunerde sequence, the shortest is the raw soil. Between the two extremes we find climax forms with AC profiles.

In regard to parent material, acid igneous rock is rare in the plains. But almost everywhere on these parallels there is a parent material that is very suitable for development studies. This is loess or sometimes loess-like sediments.

Loess is an excellent parent material for soil formation. The fact that Braunerde-Lessivé or Parabraunerde is often found in the humid temperate zone is explained by the polygenetic origin of these two soils. It is of importance that true Braunerde on volcanic and metamorphic rocks occurs in the oceanic sector, whereas Braunerde on loess is rare. On the contrary, Braunerde on loess is typical for the semicontinental sector, sometimes with Lessivé subsoil relicts that are considerably less well preserved. Loess is also an ideal parent material for Chernozem, which is formed with increasing continentality.

In Austria the border between the Braunerde province (Baltic soil province) and the Chernozem province (pannonic soil province; Kubiëna 1956c) goes right through the city of Vienna. This I was able to determine by continuous investigations of the profiles in foundation trenches and other excavations. My Viennese teacher, H. Kaserer, used to joke that Asia begins at Sanct Marx (a borough of eastern Vienna). In fact, the lowland extends from Vienna through all regions of the steppe to the desert lowland around the Aral Sea in Asia. This large area represents a unit in which the differences in climate and soil formation are in the main dictated by the degree of continentality.

A similar development caused by the same factor can be observed in the United States. This is also part of a large continent, but because of the coastal mountains, the Sierra Nevada and the Rocky Mountains, there are no lowlands in the West. The influence of the altitude of the coastal range persists on the western slope of the Rockies.

Since the Great Plains extend in a broad swathe from the Rockies a considerable distance eastward, the phenomenon is produced in reverse. The abridgement of the development sequences is produced from east to west, beginning with the ABC soils as climax formations in the central Mississippi Basin, continuing westward into Kansas, Nebraska and the Dakotas with Chernozem, Kastanozem, Burozem and Sierozem soils to semidesert. Since the main parent material for these soil developments is loess or a loess-like sediment, the soils of the Great Plains greatly resemble those of Eurasia.

INFLUENCE OF CONTINENTALITY ON SILICATE ROCKS

In Europe it is impossible to follow the influences of continentality on silicate rock in the lowland areas along the parallels between 47° and 49° N. But if we continue from the granite area of northern Spain to central Spain, we see that in the lowland or low foothill area the climax formation changes from humid Braunerde to Dry-Braunerde (mediterranean Dry-Braunerde) and brown Xero-Ranker. The mediterranean Dry-Braunerde continues into North Africa, which may be regarded as belonging to the Mediterranean zone, and then changes as it approaches the Sahara Desert into brown Xero-Ranker, and finally into a desert soil without formation of a humus horizon. This is the raw soil of the dry desert, or Yerma. In the Yerma on granite in the central Sahara (Ahaggar Mountains), the mineral grains are lightly coated with iron hydroxide, lending the sandy soil a light-brown coloration. For this reason it somewhat resembles the (B) horizon of a mediterranean Dry-Braunerde. Since an A horizon does not form, it is a raw soil—a brown Yerma.

From these observations, we can construct another idealized scheme of soil development produced by the influence of increasing continentality. This scheme is given in the lower part of Figure 11. Here the parent material is not loess but acid silicate rock.

XV

Soils of the Lowland Area of the Ever-Humid and Alternating Humid Tropics

Alexander von Humboldt was one of the first to study mountain climates, especially those of tropics, and their influence on life and morphology. One of his greatest achievements was to represent geographical facts on a tridimensional scale. In our day tridimensional presentation has been advocated particularly by the geographer Carl Troll (1955), who followed Humboldt's example in exploring the climate and vegetation of the tropical mountains.

In the higher altitudes of the tropics the soils seem very different from what we commonly think of as tropical soils. Are they indeed tropical soils? We know that they are very different from the mountain soils in the temperate and subarctic and arctic latitudes. In order to clarify our concept so that it will lend itself to tridimensional presentation, it is necessary to find out more about soil development and environmental conditions of the lowland area and the foothill belts of the tropics.

I consider the Braunlehm type of soil the most important for an understanding of tropical soil development. Its characteristics very largely explain the micromorphology and micromorphogenesis of the soils of the tropical lowland and foothill area. Braunlehm is an ABC soil that has undergone the most intense chemical weathering of any soil that exists. It is very transmutable. Thus, if we know its transformation varieties we know almost all tropical soils. It represents a final stage, far advanced from the initial (the AC) stages.

With some exceptions (mainly in the savanna areas), the initial stages of soil development are rarely found. The ratio between the existing areas of AC and ABC soils in a certain region is also an indicator of the environmental conditions of the region. Hence, the rarity of AC soils in the humid tropics is very characteristic for that region. In discussing the distribution of AC and ABC soils on limestone in different zones it was noted that nearly all soils are AC soils in the alpine and subalpine belts of the high mountains but that both AC and ABC types are present in abundance in foothill belts such as the southern Vienna Woods. In contrast, neither (A)C nor AC stages exist in the limestone region of the Lower Congo. The distribution is much the same with soils on silicate rocks.

THE SOIL MICROMORPHOLOGY OF THE TROPICS

A thorough knowledge of the micromorphology of the tropical soils is important not only for various types of investigations in the tropics but also for work in other zones. Tropical micromorphology may occur also in soils of the temperate zones, as is the case with fossil and relict soils. A knowledge of tropical soils also makes it possible to understand the formation of many polygenetic soils that developed in former geological periods and have undergone some transformation under the influence of the life and environmental conditions of the present time.

THE CONCEPT OF TROPICAL SOILS

Soils of the tropics are considered to be those which occur in the equinoctial regions, that is, those between the Tropic of Cancer and the Tropic of Capricorn, 23.5° north and south latitude.

In the corresponding tropics of the celestial globe, the sun stands at the zenith in the days of the solstices. Oblique shadows are considerably diminished. This is particularly noticeable if one photographs people in the tropics at noon. They cast almost no shadow. Peter Schlemihl, the man who incautiously sold his shadow in the story of Adelbert von Chamisso, would not have had so much trouble if he had lived in the tropics rather than in the temperate zone.

The position of the tropical sun is decisive for soil formation. In no other habitats of the earth does the soil surface receive the energy of the sun in such intense degree. This factor alone, apart from others, such as rainfall, produces extremes of climate. If soil micromorphology and soil formation depend primarily on environmental conditions and life development, then tropical soils are bound to be entirely different from the soils of the rest of the world.

In fact, the differences between tropical lowland soils and the soils of other regions are considerable. They are different even in their outer appearance, and this reflects differences in their micromorphology and micromorphological changes from horizon to horizon and from soil to soil. Since the behavior of soils is closely related to their microscopic makeup, tropical soils have microfabrics that are entirely different from those of soils of other zones. Most tropical soils are brilliant in color. In the temperate zone the soil is the most neutral in color of all landscapes. In the tropics in general it is the most striking; it may be intense yellow, ocher, orange or bright red. The soil covers, even on chemically durable parent rocks, are deep and fully weathered.

The soil properties thus developed create entirely new conditions for agriculture and forestry. These soil properties are, to a great extent, disadvantageous, for they include extremely dense structure with great instability, great erodibility, tendencies to harden, to age, and to suffer an unusually rapid decrease in fertility after cultivation. Among physical properties, the considerable plasticity and low permeability of certain sands may also be mentioned. In general, the degree of heaviness of the soil, determining whether it is difficult or easy to work with plow or spade, does not correlate with the grain size ratio (texture) of the soil.

This explains the origin of the old term Rotlehm (Wohltmann, 1892) and later its influence on the term Braunlehm as a counterpart to Roterde and Braunerde. Not the texture but the behavior of the soil for agricultural use—its heaviness—was the basis for naming it.

The sand content of a soil can be increased a great deal as a result of the formation of numerous small concretions of iron hydroxide in the soil fabric (pseudosand). Furthermore, the weight of mineral grains may increase considerably as a result of the formation of deposits of precipitated iron hydroxide upon their surface or in their fissures or cracks. This is the reason why the Heavy-Mineral Index (Schwermineralindex) of O. Tamm cannot be applied to tropical soils, although the method would otherwise be very useful as a test for the natural fertility of these soils. The method separates the "light minerals" (quartz, muscovite, orthoclase) from the "heavy minerals" (biotites, augites, hornblendes) by gravity through the application of a heavy liquid. With the precipitation of iron hydroxide on or in a mineral grain, a quartz grain may become much heavier than a tourmaline grain. All this makes it necessary to use micromorphometrical methods instead of gravimetrical analyses with tropical soils.

Special drainage techniques must be used on tropical soils that are to be cultivated. In the temperate zones the depth of drainage tile is determined to a large extent by the texture of the soil, but this rule is not valid for most tropical soils. Many sands have a very low permeability and therefore drainage tile must be laid much deeper than in the finer-textured soils in the temperate zone. Thus, for many reasons, pioneer soil experts who have gone to the tropics, after being trained in temperate regions, have found that tropical soils are very different, and they have had to develop new ways of treating them.

CLIMATIC CONDITIONS OF THE TROPICS

All the different climates of the tropics have one condition in common: sunlight and radiation are very great.

Most people of the temperate zone who want to work in the tropics are aware of the special environmental conditions and their effects and they seek a medical examination in order to determine their adaptability to the tropics. This would never be necessary for somebody from the tropics who wanted to work in a temperate region. For travelers to the tropics, vaccinations are required to protect them against diseases of the tropics. In general these are transmitted by insects or other organisms whose habitats are the very humid tropical environments. The distribution of some of these diseases coincides with the distribution of certain tropical soils.

In spite of the fact that the tropics are known as the hot zone in the equatorial region, they do not have very high temperatures. Much higher summer temperatures prevail in dry deserts like the Sahara, in latitudes of about 32° N, than in the equatorial zone. The feeling of heat in the latter is produced by the high humidity. The atmosphere is sultry.

The mean annual temperature at the thermal Equator (5° N) in Africa is 27.5°C. (81°F.). North Africa is much warmer, the mean annual temperature, in spite of the cold winter, being 30°C. (86°F.).

Very typical for the tropics are the slight temperature variations in the lowland areas; the range is only 5° to 10°C. (9° to 18°F.), and rarely that much. The figures represent the difference in temperature between the warmest and the coolest month. By contrast, in Madrid in summer, I recorded a change of 20°C. (68°F.) between day and night. This was not an exceptional but a frequent phenomenon. The small temperature range is a decisive factor in the humid tropics. It produces the hothouse climate of the equatorial rain forest. In the lowland areas of the tropics in general, climatic phenomena occur in a remarkably regular cycle. Wind direction and the change of wet and dry seasons almost follow a calendar of their own, so fixed are the dates of changing.

THE EQUATORIAL RAIN FOREST

The equatorial rain forest is an ever-humid area, with periods either richer or poorer in rainfall. Rainless periods are rare, and when they do occur, they are very short. The greatest cloudiness is observed at the thermal Equator. In fact, the more or less permanent girdle of clouds that surrounds the earth in this latitude is regarded as a characteristic phenomenon of the earth (similar to the stripes of Jupiter). The great density of the clouds leads to easy condensation and to frequent and heavy rains. The torrents that occur in this zone are matchless events. The forest roads are transformed into muddy streams that rush down the hillsides. Thunderstorms are very frequent and can be expected on about one hundred days of the year in equatorial Africa. Rarely does the sky take on a true blue color; because of the high humidity it is usually whitish. Seawater is whitish gray to leaden—in great contrast to the deep blue of the Mediterranean Sea. Wind action is weak, and there is almost never a heavy wind. This justifies the extremely light construction of the straw huts of the natives, which would be destroyed at once by gale winds. The calm zone proper is situated along the thermal Equator, the area between the trade winds. Here the velocity of the trade winds is retarded by the upward movement of the air above the thermal Equator.

Environmental conditions in different regions are reflected in the kind of life that develops, not only plant and animal but also human. Distinct races have evolved, with their own customs, habits, religions and superstitions.

Nowhere in the world did the natives of tropical regions wear clothes originally, or, if so, very little. Wearing clothes here is a kind of punishment. Because of the high humidity, clothing holds perspiration and prevents evaporation by blocking the skin from the air and air movement. The white man introduced the custom of wearing clothes to the humid tropics. But he himself knows that he has to be very careful about the kind of clothes he puts on. The fewer his clothes the healthier he will be. Too many clothes or the wrong kind cause heat stagnation and can lead to skin inflammations and rashes, and in some cases to disturbances in the circulation.

When I worked in Liberia, which has a particularly difficult coastal climate, I succumbed to five different kinds of cutaneous eruptions and had to take a sick leave and undergo special treatment when I got back to Europe. However, I was greatly satisfied with my work in Liberia, for I found exactly the kind of soils I was looking for, and in a state of perfect development. Moreover, it was an ideal location for one of my research purposes; this was to study the influence of different soils on the development of the Panama disease, which had severely attacked the banana plantations in Liberia. Thus, one might say that the bananas and I suffered from the same environmental conditions. In essence, the soils also suffer from a disease caused by the same conditions.

Among the most typical soils of that extremely humid equatorial zone, the most common and most important one is equatorial Braunlehm; also there occur tropical Pseudogley, Gley, lateritic mottled clay, lateritic Braunlehm and Braunlehm-Laterites. The last display here the best and most manifold development micromorphologically. Of particular importance is the fact that there are no tirsoid soils; instead Anmoor formations are common. All these soils have a distinctive micromorphology and form some fabric phenomena that are so stable that they may be preserved after the rest of the soil has undergone considerable transformation by environmental conditions of later periods, even by those of temperate or cold climates. These features are very valuable for the analysis of polygenetic soils, which otherwise could hardly be interpreted. Again I stress the importance of knowing the micromorphology and morphogenesis of the soils of the tropics as a basis for the investigation of many soils in the temperate and cold zones.

REGIONS OF ALTERNATING HUMID AND DRY SEASONS

Decisive for tropical regions with alternating humid and dry seasons is the fact that although there may be completely dry seasons, these alternate with seasons of great humidity. The landscape and the soils consequently undergo a marked seasonal change, a particular phenomenology in which they differ greatly from the soils of the ever-humid tropical regions. Their vegetation may be a grass savanna with gallery forests along the river banks, or forest savannas with clear and grassy deciduous forests.

With the change in plant cover, the kinds of animals also change. The fauna resembles in many aspects the fauna of the steppes. Most are plant-eating animals which survive the dry seasons by feeding on dry, hardened plant residues. We find rodents and rodent-like marsupials with rapidly growing front teeth, termites with very strong mandibles and granivorous birds with gigeria. Very typical are such gregarious animals as the gazelles, antelopes and zebras, which live in big herds. I mention these because they, like the soils, indicate the environmental conditions, and this is useful to the paleopedologist. With our present knowledge of soil formation we know that certain fossil animals and plants correlate only with certain varieties of soils. Thus, given some additional remnants of soils and soil sediments, the former soil cover can be reconstructed.

The alternation of very humid and very dry seasons has a marked effect on soil formation. Braunlehm varieties are somewhat rare, and when they do occur, it is with parent material low in iron-containing minerals and in climatic regions which do not differ very much in their seasonal change from the ever-humid tropics since they do not have pronounced dry seasons. The most typical soil types are Rotlehm and Roterde, that is, soils formed by rubification. The Roterde is also produced by red earthening. Both processes are opposite to the process of laterization.

In regions where humid and dry seasons alternate, laterization takes place only in habitats with continuously humid subsoils; the hardening of laterite layers, after they have been exposed to the air by erosion of the Rotlehm or Braunlehm surface cover, takes place more quickly and more effectively than in the ever-humid (equatorial) zone. The typical laterite varieties for the alternating humid and dry tropics are the Rotlehm-Laterites; these do not occur in the ever-humid tropics as formations of the present time.

Very typical of the tropics and subtropics with alternating humid and dry seasons are tirsoid soils, that is, formations with tirs humus. In these soils rainwater stagnates in the humid seasons, and dries up thoroughly in the dry seasons. The biology of the soils changes therefore, and they develop from anmoor-like, semiterrestric stages to strictly terrestric stages, and finally tend to mull formation.

Another very characteristic change in soil life and seasonal phenology occurs with many salt soils. Under the environmental conditions of a Gyttja (seasonal covering of the soil surface with water) they change to an Anmoor; and in the dry season, they change to a Solonchak, displaying white crusts of superficial salt efflorescences. In conjunction with dense Braunlehm and Rotlehm sediments, large areas of extremely infertile Solonetz soils can be produced.

Table 5   Possibilities of Braunlehm Transformation in the Humid and Alternating Humid and Dry Tropics

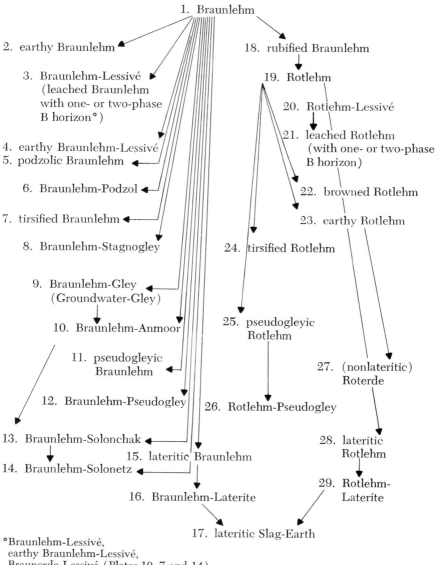

*Braunlehm-Lessivé,
earthy Braunlehm-Lessivé,
Braunerde-Lessivé (Plates 10, 7 and 14)

TRANSFORMATION OF BRAUNLEHM FABRICS

Setting aside the initial phase of soil development in the humid tropics, which has little importance in view of the intense chemical weathering, we find that almost all the soils of the advanced developmental stages have their beginning phase in a braunlehm fabric. An overview of the possibilities of transformation is given in Table 5. By the different processes of brown earthening, lessivage, podzolization, tirsification, gleyzation, pseudogleyzation, Anmoor formation, solonization, xeromorphism, rubification, red earthening, laterization, cuirass and slag-earth formation, thirty different transformation types are formed. The arrangement and combination of these types by arrow lines in Table 5 indicate their genetic interrelations.

The distinction between leached Braunlehm, podzolized Braunlehm and Braunlehm-Podzol seems to me important, since in the tropics both leaching (bleached sand formation in sandy Braunlehm) and genuine podzolization are possible. These two processes not only correspond to different habitats but represent different biological (and practical) values. For leaching as such in Braunlehm, Podzol conditions need not exist in the habitats; for the Braunlehm matrix under all conditions (even in biologically favorable habitats) is easily leached with sufficient percolation. In fact, most of the leached Braunlehm varieties represent soils of greater fertility than the unleached varieties. They are looser, have a thicker humus horizon and have excellent mull formation as a result of a well-developed earthworm fauna; the earthworms seem to favor these soils because they provide better aeration.

The processes involved in the formation of the different transformation varieties are described following Table 5.

PROCESSES INVOLVED IN THE TRANSFORMATION OF BRAUNLEHM IN THE EVER-HUMID AND ALTERNATING HUMID AND DRY LOWLAND AREAS OF THE TROPICS[*]

Brown Earthening (brown earthification, *braune Vererdung*). The process begins with brown granulation and consists of the precipitation of amorphous ferric hydroxide in the form of small amorphous granules which flocculate and form water-stable aggregates. *Nos. 2 and 4; also No. 22.* Plates 7 and 8.

[*]Figures in italics correspond to the numbers of the soil varieties in Table 5.

Leaching. Movement produced by percolating soil water by gravity. Leachable substances may move through all parts of the soil horizon (one-phase fabric) or only through conducting channels (two-phase fabric). The groundmass may consist of hardened but still peptized braunlehm fabric (Braunlehm-Lessivé) or of earthened braunlehm fabric (earthy Braunlehm-Lessivé*). The conducting channels contain peptized braunlehm fillings with fluvial fabric or more or less stratified deposition fabric. *Nos. 3 and 4; also Nos. 19, 20 and 21.* Plate 10.

Bleaching. A leaching of high intensity by which a bleached sand layer is produced. *Nos. 3 and 6.*

Podzolization. A bleaching process accompanied by dystrophic humus forms (raw humus or dystrophic moder), has a very acid reaction, and contains mostly brown humus sols in the soil solution. Frequently a B_{sh} layer is produced (never a B_h layer). In the tropics, it occurs chiefly along the banks of rivers. *No. 6.*

Pseudogleyzation. Precipitation of peptized amorphous ferric hydroxide by crystallization in the form of irregular crystallization areas in a dense groundmass which becomes more or less de-ironized. The movement to the crystallization areas is produced by diffusion. The main peptizer is colloidal silica, and in some cases perhaps humus sols as well. Sometimes some reduction of the ferric hydroxide is produced under the influence of tannic acids and acid humus sols, but it is not a decisive process and occurs primarily in transition varieties to gley formations (Stagnogley). The minerals formed by the crystallization are goethite and in some cases lepidocrocite. *Nos. 11 and 12, also Nos. 25 and 26.* Plate 11.

Tirsification. Tirs humus is a variety that occurs in alternating humid and dry areas of the tropics and subtropics. It is produced as a result of the low permeability of typical braunlehm and rotlehm fabrics. With stagnation of surface water during the humid season their biology is semiterrestric. In the dry season the soil generally dries out completely and is governed by a terrestric soil life. *Nos. 7 and 24.* Plate 5.

Rubification. Peptized amorphous iron hydroxide is crystallized in the form of tiny crystals of goethite and hematite which are freely suspended in the dense groundmass of a braunlehm fabric and give it a bright-red coloration. The crystals are visible only in the electron microscope. *Nos. 18 and 19.* Plate 9.

---

* In the temperate zone the groundmass has braunerde fabric and the conducting channels braunlehm fabric (Braunerde-Lessivé). Plate 14.

Red Earthening (red earthification, *rote Vererdung*). The flocculation of the small goethite and hematite crystals that have been produced by rubification leads to the formation of red water-stable aggregates. The soil becomes permeable. Therefore, all remaining removable substances (the peptized constituents of the soil mass) may be leached out. The first of these is colloidal silica. *Nos. 23 and 27.* Plate 9.

Laterization. Transformation by aging of all amorphous substances—mainly ferric hydroxide, aluminum hydroxide and manganese hydroxide—into crystallized deposits. Precipitation, like pseudogleyzation, is produced in the form of isolated crystallization aggregates to which the substances involved move by diffusion. The first stage results in a mottled soil mass similar to that of Pseudogley. But the support by movable substances is much greater than in the case of Pseudogley. The spots are denser, and therefore appear opaque when investigated by thin section. They grow, and finally they may coalesce into a crystalline mass which hardens irreversibly in drying. In addition to the amorphous hydroxides, colloidal silica is capable of crystallizing and serves as an effective hardening substance. But the aging and crystallization of silica take place much more slowly than with the hydroxides. Therefore many laterites have a low silica sesquioxide ratio if the silica is washed out before it can be crystallized; however, there do exist laterite layers rich in silica. *Nos. 15, 16, 17, 28 and 29.* Plate 12.

XVI

Fabric Elements of Tropical Braunlehm:
The Fabric Skeleton

Formerly, soil scientists were occupied primarily with determining the physical and chemical properties of the soil. Such analyses will always be important for soil research; however, as long as they are used alone, without consideration of the whole soil and the rôle of all its constituents in combining to form the entity of the soil, our knowledge of soils cannot but remain incomplete. This stage of soil science we might call the science of soil properties. No other natural science began with such a stage, with perhaps the exception of petrography. But petrography changed as soon as thin section analysis was introduced for the investigation of rocks; this was initiated by H. C. Sorby in 1859. As early as 1870 Vogelsang declared that no petrographer can know very much about a rock before he has seen its thin section.

Soil micromorphology reveals first of all the constituents of a fabric, and then the varieties of fabrics in a soil profile which together form the entity of the soil. The micromorphologist recognizes that these fabrics are produced by the organisms and environmental conditions in the different horizons of a soil. For the concepts of soil entity and soil individual, finding unequivocally typical series of fabrics in a set of different soil horizons is the most important question. Micromorphology implies a typical dynamics, a typical life and a typical behavior of a soil; and the soil micromorphologist seeks knowledge of the fabric elements, which are the building stones of the entity, and the building itself. The way the constituents are combined with each other to make the whole is fundamental for all micromorphological analysis.

THE FABRIC ELEMENTS

Knowledge of the fabric elements is much more important for the study of tropical soils than it is for soils of the temperate zones. It is therefore unfortunate that the science of soil properties has been incomplete for the tropical soils.

Of first importance among the soil constituents are the sand and silt fractions. The value of a soil in the tropics depends greatly on the reserve of mineral nutrients stored in its undecomposed fabric elements. Chemical weathering and consumption by the plant cover proceed much more rapidly in the humid tropics than elsewhere. For this reason, it is of great practical value to group soils by fertility levels according to (1) mineral reserves, (2) biology and (3) type of structure. It is very necessary to know the nature not only of the sand and silt fraction but also of the highly decomposed weathered products. Both are ascertained by fabric analysis.

COMPOSITION OF SOIL FABRICS

Two different parts can be distinguished in a soil fabric with regard to function: the fabric skeleton (microskeleton) and the fabric plasma. The fabric is determined by the manner of arrangement of the fabric plasma in and around the fabric skeleton.

The rôle of the fabric skeleton is that of a scaffold. It is the immobile part and characterized by a high stability of form.

The fabric plasma represents the finely dispersed part of the fabric which may become easily transportable; it is transmutable in shape and has higher or lower plasticity and tenacity than the fabric skeleton. It serves as a binding material in the soil fabric; when there is a loss of water, it is the only carrier of moisture. It contains the elements that are decisive for the activity of the soil, including the absorbing substances of cations and anions for plant nutrition.

The designation "fabric plasma" is a metaphorical expression borrowed from the protoplasm of plants because of the similarity in function. Like the protoplasm in a plant, the fabric plasma of a soil is the principal site or bearer of life, and of the physical and chemical processes. In a well-developed and fully functioning soil, it is a thickened ooze or sometimes even a jelly-like slime. It is, like the protoplasm of plants, a colloidal system containing particles of $<0.1\mu$ diameter dispersed in the soil solution.

In soils in a natural state, the fabric plasma is never without water. It dries out only in soil samples and with artificial drying. Soil samples are therefore completely different from living soils, particularly in ever-humid environments; they bear no more resemblance to the living soil than do dried plant fragments or mummified batrachians or reptiles to these organisms in nature.

The fabric plasma, like the protoplasm of plants, also contains coarsely dispersed elements in the form of splinters, granules, crystals and crystal fragments. The slimy or jelly-like matrix of the fabric plasma is produced mainly by the presence of hydrophilous colloids. This plasma may contain large or small amounts of flocculated elements, or none at all. Or it may be completely flocculated to a sponge-like substance.

Crystals may be formed from a solution or by crystallization of peptized amorphous substances (such as iron hydroxides transformed by rubification, pseudogleyzation and laterization). Crystallization of salts from genuine solutions in the fabric plasma can be so strong that a bursting and mechanical decay of the soil fabric are produced. With other conditions, a complete hardening by cementation of the soil mass can take place. The salts involved in these processes are, first, calcium carbonate, calcium sulfate and calcium chloride, and then the alkali salts. When the fabric plasma is hardened as a result of compacted crystallization of amorphous substances, a considerable decrease in the biological activity of the soil may take place. This process is therefore a typical form of soil aging.

In the natural soil a liquid phase of the soil complex can be observed in addition to the fabric skeleton and the fabric plasma: this is the soil solution. It can be investigated only by direct microscopy in undisturbed soils in situ or in fresh frame samples extracted from the natural environment. Knowledge of the arrangement of the soil solution in the soil fabric at different stages of saturation is of great importance for various fields of soil research. It is an essential part of the science of soil micromorphology.

Between soil and life there exist close interrelations. Living bodies are colloidal systems. We know also that the fabric plasma is, in essence, a colloidal system, although it consists to a great extent of crystalline inorganic constituents. Because of this, the fabric plasma may be regarded as the most primitive colloidal system, in other words as the real protoplasm—the system that initiates the transformation of inanimate matter into vital matter.

The boundary between the fabric plasma and the soil solution is very indistinct in soils in situ. This is the reason why thin sections of some semiterrestric and subaqueous soils are difficult to prepare satisfactorily. Examples are the different forms of Gyttja, particularly Peat-Gyttja. In these soils, the fabric plasma contains so much water that when the sample is dried it shrinks to a tenth or less of its original volume. The thin section does not, in this case, retain the real aspect of the fabric. We see from this that the ratios between the fabric plasma and the soil solution are very important.

The same is true for the ratios between the fabric skeleton and the fabric plasma. These ratios should be more or less balanced in all soils. To draw a parallel with the human body, we might compare the fabric skeleton with the bones and the fabric plasma with the fleshy connective tissue. Men with too much skeleton, who seem to be all skin and bones, and those with too much "plasma" both have inferior constitutions. An ideal ratio lies, of course, somewhere between the two. With soils, the most favorable ratio varies with the soil type. A small soil skeleton is especially disadvantageous in many tropical soils, for it reflects unfavorable water and air conditions. The nature of a soil fabric is determined by the way the fabric skeleton is combined with the fabric plasma. Deficiencies in both fabric plasma and fabric skeleton induce the worst biological conditions in natural soils.

When I started with fabric analysis some forty years ago, I tried to follow the example of petrography by giving each fabric type an independent Greek or Latin name; they are all listed and described in my first book *Micropedology* (Kubiëna, 1938). From "agglomeratic" and "intertextic" I continued to "chlamydomorphic," "magmoidic," "porphyropeptic" and "porphyropectic" fabrics. But gradually I found micromorphology so complex that by this system I would need quantities of new names—so many that I felt that the micromorphological terminology would become unnecessarily difficult for the soil scientist who was not a specialist in soil fabrics. For me, micromorphology has always offered a way to arrive at a closer understanding of the soil as an entity. It gave me much satisfaction to verify, by comparative research, that every well-developed soil type has a micromorphology which characterizes it unequivocally. I began therefore to designate soil fabrics by the name of the soil type or the humus form. By doing this I wanted both to avoid a separate vocabulary and to indicate the soil type or humus form to which the particular fabric belonged. Here I must mention a scientist who adopted my first system. He is Roy Brewer, who has written an extensive and exhaustive book on the subject (1964). I follow his endeavors with interest; I only hope that his nomenclature will not make the understanding of micromorphological descriptions too difficult.

QUARTZ IN THE FABRIC SKELETON OF TROPICAL SOILS

It is of great significance that even single fabric elements may have characteristics which make it possible to recognize the soil type in which they originated or were transformed. A single mineral grain or a tiny fragment of the fabric plasma may be sufficient for identification.

In the humid tropics, quartz is the mineral that occurs most frequently in the sand and silt fractions; in many cases it is the only mineral present. Quartz is not a ballast in the tropical soil but a highly useful mineral, although not a source for plant nutrition. It is the mineral most resistant to weathering. In temperate climates it is nearly insoluble. But in the humid tropics it has some solubility and is transformed into colloidal silica. This is often recognizable by the grain forms. Old laterite crusts may lose their quartz grains completely and be transformed into quartz-free formations that are called "rich ores" (*Reicherze*) by prospectors. The primary and secondary quartz of these old (mostly Cretaceous) laterite layers is completely dissolved and washed out.

Since quartz is the principal constituent of the fabric plasma, quartz grains are easily recognized. In humid soils the quartz grains retain their intense luster. Their surface sparkles. In thin section, quartz is easy to recognize by its complete lack of cleavability, its irregular to conchoidal fracture, and its fairly low double refraction which, at the standard thickness of the preparation (0.03 mm.), is steel gray to whitish and bright yellow in color. It is rarely absent in tropical soils, and may occur even in soils produced from basic parent rocks that are practically free of quartz—such as the basalts, gabbros, dolerites, norites and diorites.

The quartz that originates in these soils is secondary quartz produced by the crystallization of aged colloidal silica. This phenomenon is particularly common in lateritic soils. The secondary quartz has a kind of chert structure; a black and white spotted pattern throughout its interior can be seen with crossed Nicols. Such secondary quartz, together with iron hydroxide concretions, is frequently the main constituent of the so-called "stone line," which often occurs in profiles of soil sediments of the African tropics. It indicates former soil surfaces that were exposed to a dry or semiarid climate. It was produced by sheet erosion of an almost bare soil surface lacking a closed protective vegetation. The erosion removed the finely dispersed constituents and left the gravels behind on the newly formed surface.

Quartz grains in tropical soils often have a pronounced rounded surface. Under humid tropical conditions, a morphology similar to that of
a karren formation on limestone can be produced on the grain surface.
Fillings or depositions in dents or cracks of the grains are very typical
of tropical soils. They may be: (1) aged residues of braunlehm plasma.
(2) deep-brown opaque amorphous concretionary fillings of iron hydroxide (these iron hydroxide precipitations indicate that they originated
in ordinary braunlehm fabrics). (3) opaque red recrystallized concretionary iron hydroxide, mostly hematite or goethite, produced by rubification, pseudogleyzation or laterization. Vertical or incident light is
needed to investigate them properly. (4) aged residues of braunlehm
plasma in the cracks, rubified or red earth deposits on the grain surface
(Braunlehm remnants in Rotlehm). (5) opaque red crystallized concretionary iron hydroxide in cracks, or red earth deposits on the grain
surface (from rotlehm fabric).

THE FELDSPARS

The feldspars represent the most important parent materials for clay
formation. They are not present in most soils of the humid tropics because
of the intense weathering, but where they are present they are a good
indicator of fertility. Not only are they sources of potassium, sodium and
calcium, but their very presence makes it probable that other minerals
will also be present, serving as additional carriers of plant nutrients. The
events that lead to the presence of unweathered feldspars in tropical
soils are usually as follows.

We know that soils occur not only as formations in situ but also in
the form of soil sediments. If the soil sediments are materials that have
been washed down from mountain regions into lowland areas, they will
have an excellent earthy structure of high stability and they will prove
of great benefit to the agriculture and forestry of the lowlands. This is
particularly true in lowland regions where soil formations consist only
of very dense, highly weathered Braunlehm and Rotlehm varieties with
poor soil life and scanty humus formation which are consequently low
in fertility. The favorable structure of the sedimentary materials is produced by environmental conditions in the uplands, particularly by the
diurnal change of temperature which causes earthy Braunlehm to be
formed. In the process of erosion, the earthy Braunlehm material is deposited in the lowlands along the river banks and becomes new soil in the

new environments since it is subject to new and, in most places, very favorable humus formation. We call these new soils earthy Braunlehm-Vegas, Vega being the type name of the soils of the river valleys.

Besides the earthy Braunlehm sediments, some other important material can be transported to the lowlands in the process of erosion. If the upland soil profile in situ is completely eroded by heavy rains, the (B)/C and C horizons are washed away, and with them many unweathered minerals, including fresh feldspars. Where such minerals are included in the soil sediments, the result is the formation of the most fertile of all tropical soils.

In regions where the parent rocks of the uplands are rich in bases, the earthy Braunlehm-Vegas are characterized by an excellent water-stable structure, favorable water conditions, intense soil life and deep humus formation, with high nutrient reserves. Such soils occur both in the ever-humid tropics and in regions of alternating humidity.

In Colombia these soils form the famous banana district at the foot of the Sierra Nevada of Santa Marta. The excellent mull horizons with well-developed sponge fabric reach a thickness of 63 to 76 cm. In the course of comparative investigations on the effect of soil conditions on the development of the Panama disease in the banana plantations, we found that these soils together with some earthy Braunlehm varieties in situ were the most resistant to the spread of the epidemic.

Feldspars and other easily weathered silicates may also be present under certain conditions in lateritic soils. In the process of laterization, amorphous iron hydroxide is precipitated, not in the form of tiny isolated crystals embedded in the soil matrix (as with rubification), but rather in the form of coherent crystal aggregates. They may in the end intersperse the whole fabric (causing induration after drying out). Or they may fill out cracks in the soil or fissures in cleavable minerals like feldspars. Or the surface of the grains or fragments may become so coated that they are entirely encased in very dense and opaque crusts of crystallized iron hydroxide; this encasing isolates the minerals and protects them against the very active weathering agents of the tropics. The occurrence of this latter phenomenon was not known until soil samples were subjected to thin section analysis; however, by this method, the feldspars, encased and intersected by iron hydroxide, are easily recognized as derived from laterite layers.

The feldspars are readily identified in the thin section by their high cleavage and the low double refraction, which, at standard thickness, shows interference colors from steel gray to light gray. Fresh feldspars have a glass-like luster; but in the soil or in an even slightly weathered rock, they are dull and have no luster at all because they are in the process of decomposing. They are thus easily distinguished from the quartz grains, which retain their intense fat-like luster.

The potassium feldspars, with a composition of $KAlSi_3O_8$, represent the most important potassium sources of the soil. Microcline, the triclinic variety, can easily be distinguished under crossed Nicols from orthoclase, the monoclinic variety, because of the characteristic spindle-shaped twin lamellae which cross each other at right angles (gridiron or quadrille structure). In temperate regions microcline is regarded as somewhat more difficult to weather than orthoclase; but in the tropics there is not very much difference between them in this respect.

Among the triclinic plagioclases are some very important sources of calcium. Since calcium is rare in humid tropical soils, the presence of calcium reserves is of great value. In the following sequence, the plagioclases are listed according to their sodium or calcium content.

Albite	$NaAlSi_3O_8$
Oligoclase	Ab_8An_2
Andesine	Ab_5An_5
Bytownite	Ab_2An_8
Anorthite	$CaAl_2Si_2O_8$

Ab = Albite
An = Anorthite

The plagioclases generally can be easily distinguished in thin section from the more common potassium feldspars because of their characteristic twin lamellae. The wider the lamellae, the richer a plagioclase is in calcium. Plagioclases ordinarily are not found in tropical soils because they decompose readily. They, as well as the potassium feldspars, are rapidly transformed into clay minerals, primarily into halloysite, fireclay minerals, kaolinite or gibbsite, according to the environmental conditions.

The manner of weathering is different. In the temperate zone the weathering is visibly produced at the surfaces or in the cleavages of the mineral grains. In Podzol soils it is accompanied by the formation of weathered cavities in the fissures of the minerals. Their walls are covered with brown coatings of acids which play an important rôle in the weathering of the feldspars. In tropical soils a complete softening of the whole mineral substance is produced—not only with feldspars but with most of the other silicates. With parent rocks low in quartz in the (B)/C and the C_1 horizons, the structure of the former rock, including the crystallographic delimitation of the single minerals, is completely preserved; but the whole mass of the horizon is easily cut with a knife. The feldspars and other silicates are transformed into clay minerals, with perfect pseudomorphosis extending through the whole horizon.

This is also the case with the fabric of gabbros, norites and diorites. Because of the density of the fabric, penetration of oxygen during weathering is greatly impeded and the decomposition takes place without oxidation. In this case, the horizon does not become brown, but remains gray. We designate horizons of that kind as C_m horizons (m = massive). They occur frequently in extremely humid soils. In spite of the gray color, they are not gley horizons, because there was no reduction but simply a lack of oxidation in the weathering process. They represent a form of fragipan. A fragipan can be of (B)/C, B/C, E/(B), E/B, G/C, g/C or G character. With the C_m horizon, a new form is added.

Another mode of weathering takes place if an isolated fresh feldspar fragment is embedded in a Braunlehm sediment in a tropical humid environment.

A kind of karren formation is produced on exposed hard rock in the tropics, not only on quartzite or other siliceous rock but also on silicate rocks such as gneiss and granite. That kind of transformation on the rock surface may be in part the result of a solution weathering, in part the result of a rapid chemical decomposition with intense leaching of the weathered products. This transformation is much more intense at the surface of a feldspar fragment if the fragment is embedded in a humid braunlehm matrix in a tropical environment.

THE MICAS

Like the feldspars, the micas are not usual constituents of humid tropical soils. Even muscovite, which is fairly resistant to weathering in temperate climates, completely decomposes in the ever-humid tropics. Thus micas in the parent material are important sources of potassium.

A large biotite content in the parent rock greatly favors two important processes which contribute to the improvement of the very dense structure of many tropical soils. These processes are (1) brown earthening (*braune Vererdung*) and (2) rubification, followed by red earthening (*rote Vererdung*). The biotite has this influence in common with other minerals which yield a large amount of free iron hydroxide in the course of decomposition, for example the chlorites, olivines, augites and hornblendes.

Whereas the muscovites have a relatively simple chemical composition —$HKAl_3Si_3O_{12}$—the biotites contain in addition to this a component of olivine of varying composition. Olivine—$((MgFe)_2SiO_4)$—is greatly subject to decomposition by weathering. Therefore, the dark-green, olive-green to blackish-green biotites change color rapidly with the start of weathering and turn to brown, reddish, yellow and golden tints, or become colorless if freed iron moves out of the crystals.

Both micas are transformed into clay minerals. Although in temperate and subtropical environments muscovite forms primarily illite, and biotite forms vermiculite—both of which, in fabric plasmas, show very marked double refraction by particle arrangement in the form of streaks and bands—in the tropics the weathering continues to halloysite, fireclay mineral or kaolinite.

The micas are easily recognized in thin section because they occur in laminated plates and have high interference colors; biotites have, in addition, a marked pleochroism if cut across the laminae. Biotites have this property in common with chlorites. Chlorites, however, are colored differently and frequently show positive optical character, whereas biotite is always negative.

THE AMPHIBOLES AND PYROXENES

Both the amphiboles and the pyroxenes are important subgroups in the parent rocks of tropical soils, but they rarely occur in the soils themselves. Both are metasilicates of magnesium, calcium, iron, manganese and aluminum of different composition. A content of alkalies is also possible with the amphiboles but not with the pyroxenes. Amphiboles and pyroxenes are carriers or indicators of soil fertility (Ca, Mg, alkalies, phosphoric acid). They are also sources of clay. They yield, upon weathering, considerable amounts of iron hydroxides which play an important rôle in producing different types of soil structure.

Rotlehm and Roterde are often found on basalts, gabbros and norites in the tropics; of course, the climate must be one of alternating humid and dry seasons. However, the rubification of soils on these parent rocks is produced more easily also in regions which are situated relatively close to the ever-humid zone. The opposite is true with parent rocks low in augites and other iron hydroxide-producing minerals, such as granites and acid gneisses. Granites and acid gneisses generally need a climate with very long and very effective dry seasons for rubification to take place. The latter are quite often associated with relict soils in Spain and in the central and southern Sahara.

It is of particular interest that the above-mentioned parent rocks behave differently with laterization. Laterites with a well-developed micromorphology occur much more frequently on acid parent rocks with a low content of basic minerals which develop abundantly free iron hydroxide by weathering. I was able to demonstrate this in 1955 at a meeting of CRACCUS (Comité Regional d'Afrique Centrale pour la Conservation et Utilization du Sol) at Santa Isabel, on the island of Fernando Póo (Gulf of Guinea). In the basaltic regions, particularly the tropical East Atlantic islands, there was almost no laterization but there was intense rubification on the border of the ever-humid equatorial rain forest. However, in the gneiss region of Río Muni, laterization was very general.

By comparative micromorphological investigation of laterites of different genetic stages, one can observe that the formation of perfect laterite fabrics depends on slow laterization over a long period of time and on the presence of highly peptized and movable amorphous substances which are able to migrate by diffusion over long distances. (These migrations remind one of the so-called geological diffusions in rocks.) The decisive substances for the formation of the framework of laterite fabrics are, in the final stages, iron hydroxides.

In layers of soils or soil sediments rich in free iron hydroxides, iron-bearing compounds are easily precipitated. This greatly decreases their mobility, and in consequence a further development of the highly compact and completely opaque aggregates by crystallization of the amorphous iron hydroxides essential for the lateritic framework is not possible. Although pyroxenes and amphiboles are rarely found in tropical soils of the lowland areas, it is evident from the foregoing that they play an important rôle as parent materials and hence a decisive one in determining the kind of soil formation. Even in the equatorial rain forest, the Braunlehm varieties on basalt rich in augites differ considerably in character from the ordinary equatorial Braunlehm. They are darker in color, owing partly to the higher content of iron hydroxide, and partly to the content of almost undecomposed blackish magnetite crystals whose crystallographic delimitation (primarily fragments of cubes) is well preserved.[*]

Pyroxenes and amphiboles are in general easily recognized in thin section. They have much higher interference colors than quartz and the feldspars and, at standard thicknesses, are red and violet to blue-violet. The amphiboles display in addition a very marked pleochroism, which is little developed with the pyroxenes or, in many cases, lacking entirely. Cleavability is good in both, in fact excellent in the amphiboles. Both occur primarily in the form of small columns or grains. If cross sections of columnar fractions are available, the pyroxenes can be distinguished from the amphiboles by their eight-sided cross section. The amphiboles have six-sided cross sections. The cleavages in these cross sections of the pyroxenes occur at an angle of about 90 degrees; the amphiboles at an angle of 124 degrees.

Both groups of minerals can be divided into three divisions according to their crystal system. Both have rhombic, monoclinic and triclinic varieties. The triclinic varieties have little importance for soil formation. Most important are the monoclinic varieties, particularly those which contain aluminum oxide in addition to iron. They are called augites in the group of pyroxenes and hornblendes in the group of the amphiboles. Common augite occurs in basalts, and the corresponding hornblende is the basaltic hornblende.

Diopside, a monoclinic pyroxene variety free of aluminum is important for tropical soil formation because it contains calcium; diopside has a composition of $CaMg(SiO_3)_2$. There also exists a form of the same mineral which is rich in iron and contains some aluminum oxide. This is diallag, which occurs in diabase and gabbro. Of some importance for soil formation are the rhombic pyroxenes. Their main forms are as follows:

Enstatite	$MgSiO_3$; is light in color and low in ferrous oxide.
Bronzite	has 5 to 15 per cent ferrous oxide, is a luster-like bronze and has laminated aggregates.
Hypersthene	$FeSiO_3$; has 15 to 30 per cent ferrous oxide, is similar to bronzite, may also have foliated aggregates but is darker in color.

---

[*] Undestroyed magnetite crystals occur even in rotlehm fabrics. Only in highly earthened roterde fabrics do they burst and become bright red in color. Evidently the magnetite crystals are transformed into maghemite. R. C. Mackenzie of Aberdeen verified the transformation of magnetite to maghemite in a Rotlehm sample.

These minerals occur frequently in the gabbros and norites, rocks which are easily weathered. Without oxidation, as in C_m horizons, they do not produce free ferric hydroxides and lend only grayish colors to the soil mass. In the C_m horizons the rock structure is completely preserved, but the mass has softened and can be cut with a knife.

The pyroxenes weather considerably more quickly than the amphiboles, but the amphiboles are not found in well-developed tropical lowland soils.

XVII

Fabric Elements of Tropical Braunlehm: The Braunlehm Matrix

The Braunlehm of the equatorial lowland region is a unique formation with a weathering intensity that has no counterpart on earth, even in other ever-humid tropical climates. Braunlehm occurs in all continuously humid regions of the tropics and subtropics and may be found also in the alternating humid and dry zone if for some reason rubification is impeded (usually because of insufficient iron content). But the most typical Braunlehm, with the most characteristic matrix, occurs in the ever-humid equatorial zone.

The most striking characteristic of this matrix is that it contains a very large amount of amorphous and highly peptized weathered products in the form of a very dense mass which is subject to great swelling and shrinking and which is easily silted up. Because of these properties it has a high binding power when in a dry and moderately moist state, but as its water content increases it is transformed into a formless mud. All the structures in this matrix are very unstable, whether because of mechanical loosening or because of the natural aggregate formation that occurs as a result of the shrinking of the matrix. When the matrix is silted up in streams or pools or puddles, the water turns into a muddy yellow liquid (like pea soup) which clears up somewhat only after a long time.

Under the microscope, the braunlehm matrix appears as a dense clayish mass of great uniformity and a vivid egg-yellow color (yellow 2.5 Y 8/8 to 5 Y 8/8). Seen under reflected light with a magnification of 80 times or more, it has a slight luster, like wax.

The main constituents of the braunlehm matrix are clay minerals, hydrated colloidal silica, amorphous and peptized ferric hydroxide, amorphous aluminum hydroxide and sometimes manganese hydroxide. Subsidiary constituents may be crystalline iron or aluminum hydroxides; however, these originate as a rule in some pre-stage of a polygenetic soil formation.

THE CLAY MINERALS

The clay substances of the matrix of equatorial Braunlehm, that is, the hydrated alumino-silicates, are the only constituents of this soil matrix that are crystalline rather than amorphous. There is another matrix very similar to the braunlehm matrix that is produced not by weathering but by postvolcanic influence. Humidity in conjunction with high temperatures can produce such a matrix; when it is produced in the extremely high temperatures of a humid tropical climate, the clay substance may also be completely amorphous, taking the form of allophane varieties. However, clay minerals of the simplest construction, and highly resistant to further transformation, are finally accumulated in the matrix of equatorial Braunlehm; most frequently found are halloysite and fireclay minerals. There may occasionally be other clay minerals, but they are rare and occur because of circumstances that alter the influence of the typical environmental conditions.

Halloysite belongs to the kaolinite group of clay minerals; it is a two-layer mineral, with one silica oxide tetrahedron layer and one aluminum hydroxide octahedron layer with an electrically neutral surface and a nonswelling lattice. As in kaolinite, the distance between the layers is unchangeable. The composition of halloysite is $Al_2O_3 \cdot 2SiO_2 \cdot 4H_2O$. The crystals are six-sided plates. They are monoclinic (pseudohexagonal), and even the largest ones measure less than 1μ. They differ from kaolinite in that they form little rolls, which evidently consist of rolled packages of kaolinite crystals with intermediate layers of water, because with dehydration the rolls open and split apart. Such a dehydrated halloysite, called metahalloysite, has a composition of $Al_2O_3 \cdot 2SiO_2 \cdot 2H_2O$. It can be produced by heating, but in the tropics it occurs in aged Braunlehm, that is, in transition varieties to lateritic soils and in Laterites.

Halloysite has very low double refraction or no birefringence at all. Birefringence is a little higher if particle arrangement takes place, but this mineral never displays the peculiar birefringent streaks and coatings that are produced by particle arrangement in illite- or vermiculite-rich layers. This difference in behavior with regard to particle arrangement and double refraction makes it possible to use the polarizing microscope to ascertain at least partially the character of the clay minerals present in thin sections. This process is facilitated by the use of phase-contrast methods (Altemüller, 1964). A fairly low birefringence is also typical of kaolinite crystals, which likewise are monoclinic (pseudohexagonal, forming six-sided plates); particle arrangement produces little development of the double-refractive streaks and coatings which are quite common in Braunlehm and Rotlehm varieties of the subtropics.

Another mineral that occurs frequently in the matrix of equatorial Braunlehm is fireclay mineral. It is very similar to halloysite and kaolinite but exhibits some irregularities in the construction of the lattices. The crystals are also six-sided plates, smaller than those of kaolinite (less than 0.5μ).

Kaolinite with a composition of $Al_2O_3 \cdot 2SiO_2 \cdot 2H_2O$ or $Al_2Si_2O_5(OH)_4$ (only half the water content of halloysite) is the most common clay mineral of the alternating humid and dry zone of the tropics, particularly in Rotlehm and Roterde varieties. Halloysite and kaolinite have a low swelling capacity, and therefore behave differently from, for instance, the clay minerals of the montmorillonite group.

It seems paradoxical that the halloysite matrix of equatorial Braunlehm has the greatest swelling capacity, adhesiveness and plasticity of all soil fabric plasmas. Largely responsible for these qualities are the very exceptional properties of the Braunlehm soils—extremely dense structure with little stability, poor permeability, low air content, great tenacity in a dry state, and the ability to transform into a homogeneous mud with high water content. The reasons for these properties lie in the fact that the braunlehm matrix contains not only clay but also such other constituents as hydrated colloidal silica, amorphous ferric hydroxide and aluminum hydroxide.

HYDRATED COLLOIDAL SILICA

This constituent is of special importance for the braunlehm matrix. Its condition, its degree of aging and its peptization power are decisive for the character of the whole soil. Colloidal silica is a hydrophilous colloid of great stability. Of all colloids in the soil, it is the most effective peptizer. Its very low tendency for atomic ordination is preserved with complete drying; so also is its great absorption capacity. In respect to its low tendency for atomic ordering, colloidal silica is similar to the flocculated amorphous ferric hydroxide of the temperate zone, which has remarkable stability as an amorphous colloid but not nearly as much as colloidal silica. Whereas flocculated amorphous ferric hydroxide can be easily transformed into crystalline hematite by heating, amorphous silica retains its amorphous nature even when ignited. By this process a very light powder of amorphous silica dioxide is produced.

This compound occurs also in nature in the form of inclusions in volcanic rocks, and is known as lechatelierite (after the mineralogist Le Châtelier, 1913). It also occurs very characteristically in the form of sand tubes, produced when lightning strikes into a sand layer and causes the quartz grains of the sand to melt and transform from a crystalline to an amorphous state. The mineral substance produced is fulgurite. I well remember seeing fulgurites in the desert sand of the Great Western Erg in the Sahara.

Soils with a large amount of braunlehm matrix, which ordinarily are exposed to a warm, dry climate, can harden so completely that they lose their former swelling capacity and are not subject to erosion. They harden like rock, largely because the hydrated colloidal silica is transformed into opal. Transformations of this kind are common in dry subtropical and mediterranean climates and in Braunlehm relicts in dry regions of the United States.

Colloidal silica is colorless, and therefore it is not visible in thin sections of soils. Even an opalized silica is rarely recognizable in the fabric. But after advanced aging, the silica crystallizes and becomes visible because of its double refraction. This crystallization takes place particularly in laterized braunlehm fabrics if, for some reason, the silica was not washed out while it was in a hydrated, highly leachable state. Hydrated colloidal silica is a very effective protecting colloid not only for ferric and manganese hydroxide but also for humus substances that are found on the mineral decomposition products.

A distinctive phenomenon in a humid tropical climate is silicification. Braunlehm layers on limestone (tropical Terra fusca) very frequently impregnate the upper part of the limestone rock with hydrated colloidal silica. Transformed by aging into opal, it forms an insoluble new fabric pattern of its own. The soluble calcium carbonate under the living soil cover gradually moves out from the humid rock. With continuous aging, the silica crystallizes and is transformed into secondary quartz. This quartz, which is readily visible with crossed Nicols under the polarizing microscope, exhibits the characteristic chert structure. Since the silica is colorless, the bare rock exposed to the air after removal of the Terra fusca cover still has the appearance of a limestone to the naked eye. But testing with a drop of dilute hydrochloric acid proves that all calcium carbonate has been removed.

Particularly evident in this alteration process are the changes in the small fragments of limestone that become embedded in the braunlehm matrix of a tropical Terra fusca. All these fragments silicify and decalcify, and are transformed finally into secondary quartz. The delimitation of the limestone fragments can still be discerned, but the characteristic interference colors of quartz and its chert structure can be seen in a thin section of the interior of the Braunlehm.

AMORPHOUS FERRIC HYDROXIDE

In thin section analysis, this very important and characterizing constituent of most tropical soils is easily recognized by its yellow color in the peptized state, or its brown to dark-brown color in the flocculated state (earthy Braunlehm). Amorphous ferric hydroxide is difficult to see by the usual x-ray methods, but since all iron hydroxides are transformed by heating into crystalline hematite, the amorphous type can be recognized by comparing nonignited and the ignited samples of the matrix.

Peptized amorphous iron hydroxide present in the braunlehm matrix is relatively easily transformed by crystallization. With higher temperatures, this transformation is considerably accelerated. It is transformed into hematite within a few days at a temperature of 90°C. At room temperature it may take several years—still a short time in terms of the time necessary for changes in soil development. This is one of the reasons why amorphous iron hydroxide is easily crystallized in tropical lowland regions, whereas it is very stable in the temperate and cold zones of the globe.

Amorphous iron hydroxide is irreversibly shrunk by intense drying. However, complete shrinking does not change the atomic order, no crystallization is produced and the amorphous stage is stabilized considerably.

ALUMINUM HYDROXIDE

Aluminum hydroxide also is often a constituent in the matrix of equatorial Braunlehm, although it is not in much evidence because not only is it colorless but it is considerably peptized. Moreover, it does not play any particular rôle in the matrix. But as soon as Braunlehm undergoes laterization, aluminum hydroxide may be accumulated, partly purified of its iron hydroxide content, and finally crystallized. The main minerals formed by this process are gibbsite ($Al_2O_3 \cdot 3H_2O$) and böhmite ($Al_2O_3 \cdot H_2O$).

XVIII

Laterite Formation

F. Buchanan (1807) gave the name laterite to a kind of clay that contains great quantities of iron hydroxide in the form of red and yellow stains and which is soft enough to be easily cut when in the fresh state but hardens when exposed to the air and thereafter weathers only with great difficulty. This first description of the final product of laterization is unequivocal, and much that was added in later descriptions of the formation unsuitably narrowed the concept of Buchanan or broadened it incorrectly.

Somewhat like pseudogleyzation, laterization is produced as peptized amorphous substances are precipitated by crystallization in more or less compact crystallization complexes, but, in contrast to pseudogleyzation, all the space between these complexes is gradually filled up and they grow together into a crystalline mass which, with desiccation, hardens to a rock-like consistency.

According to this definition, which can be verified by looking at the peculiar lateritic micromorphology, Laterites and lateritic soils must be distinguished not only from Pseudogley but also from soils produced by rubification (Rotlehm, earthy Rotlehm, Roterde, Terra rossa) and from Gley and Raseneisenstein (bog iron formation).

Laterization as a pedogenetic phenomenon can take place only in certain soil fabrics which conduct peptized amorphous substances by diffusion (tropical braunlehm and rotlehm fabrics).

In view of the results of soil development research (determination of development phases and their genetic order) and of comparative investigation of habitats and of comparative micromorphological analysis, the following facts should be emphasized:

(1) There is no one "laterite climate" as such that determines the production of Laterites and lateritic soils. These soils are produced both in the ever-humid and alternating humid and dry tropics, but only in suitable habitats. These are not habitats of great water permeability but,

on the contrary, dense loamy to clayish soil covers or sediment layers in which the more or less constant moisture conditions and small temperature changes permit a constant movement of the peptized amorphous substances towards their crystallization centers.

(2) When laterization is produced in the zone of the ever-humid equatorial rain forest, laterite fabrics undergo an even more continuous and therefore more perfect development. The formation of completely dehydrated oxides like hematite does not require desilicification of the mother substance. It can be produced by aging in ever-humid environments.

(3) Laterite layers harden irreversibly when desiccated. An alternating humid and dry climate with long rainless periods does not favor laterization as such, but only the secondary hardening of laterites to cuirass and slag layers. These become surface formations only if the nonlaterized cover (of Braunlehm or Rotlehm) is removed by erosion.

(4) A laterite layer becomes permeable and open to percolation only when a water-stable filter scaffold in the form of a coherent crystalline iron sponge within the dense matrix is produced.

(5) Low content of colloidal silica is a common but not a necessary requirement for laterite formation. If laterization is not disturbed by a premature leaching, the silica also remains in the fabric. Silica-rich Laterites are thus produced in which the silica becomes transformed after some opalization into a crystalline state, forming secondary quartz varieties which have a chert structure. Young laterite horizons (mottled clays) are, as a rule, rich in silica but nevertheless must be regarded as typical products of laterization. The degree of decrease in silica cannot be regarded as a measure of the degree of laterization but only as a measure of the intensity of leaching of colloidal silica in laterite layers before the silica could become stabilized by opalization or crystallization.

(6) Laterization and leaching of laterite layers are two different processes whose effects are easily distinguished by micromorphological investigations. In essence, they replace each other. With the loss of the mother substrate, the matrix of the braunlehm or rotlehm fabric, the laterite layer loses the ability to undergo further constructive development; it petrifies in some intermediate stage and is thereafter subject to the removal of only a part of its initial fabric constituents.

(7) Pisolite formation (accumulation of pea-shaped iron hydroxide concretions) is not a characteristic of laterization, nor is it produced by a

"laterite climate." It belongs to the normal dynamics of tropical and even subtropical Braunlehm varieties. Unusually dense iron concretions can be produced by accumulation as a result of erosion of surface layers of Braunlehm on slopes or by a local concentration of peptized iron hydroxide, both of which processes frequently can be regarded as a pre-stage of laterization.

(8) De-ironized clayey white horizons (pallid zones) occur with lateritic soils but may also be formed as a part of nonlaterized Rotlehm profiles (C_w horizons). They are particularly common in the Tertiary formations of the Mediterranean and the Sahara.

(9) True Laterites often display numerous traces of a former terrestric soil life, particularly in tubes and spaces produced by earthworms or termites and generally filled out with soil material younger than that of the groundmass. Where Laterites can be confused with ironstones of volcanothermic origin or bog iron formation, they are good indicators of genuine laterization.

(10) Laterites and lateritic soils may be common in some regions, but in general they are less common than nonlateritic tropical soils, such as those produced by rubification (Rotlehm and Roterde varieties), or by brown earthening (earthy Braunlehm), or by pseudogleyzation, or those which represent unchanged Braunerde varieties. These nonlateritic soils correspond to environmental conditions which differ from those which influence laterization as well as from those which make them different from each other. From the practical viewpoint and for comparative environmental research and soil systematics, these soils, which have a different genesis, profile morphology, micromorphology, behavior and practical value, should be distinguished from the genuine Laterites and lateritic soils. To identify them, even in the initial stages, there is no more adequate method than micromorphological analysis.

(11) Eluviation and decrease of colloidal silica occur not only in Laterite but also in earthy Braunlehm, earthy Rotlehm and nonlateritic Roterde. By chemical analysis alone, therefore, it is not possible to distinguish lateritic from nonlateritic soil formations.

(12) Edaphoidic ironstones and mottled clays are very frequently confused with Laterites and lateritic soils. The former are produced by postvolcanic influences and have a soil-like appearance, but they are not soil formations. Conclusions concerning a lateritic genesis in volcanic areas must therefore be established with care.

DEVELOPMENT OF THE LATERITE PROFILE

Lateritic profile development differs from profile development produced by rubification. With rubification, the transformation begins with the surface layer (climatic influence in alternating humid and dry environments) and proceeds into the interior of the profile. Often the (B)/C horizon still has the yellowish color of the nonrubified Braunlehm. The normal succession of the horizons is therefore A, (B), (B)/C, and C (see Figure 12).

In lateritic profiles the transformation is produced in the interior soil in layers which are subject to little change in humidity and temperature during the course of the year. Undisturbed but considerably aged Laterite profiles display a nonlaterized surface cover with braunlehm or rotlehm fabric (see Figure 12), followed by a cuirass layer, sometimes of considerable thickness, over a mottled zone of reddish spots in a whitish groundmass, gradually merging with the more or less fresh parent rock. The decay layer above the fresh C horizon (parent rock) may have a different character. It may be a kind of C_m horizon that has undergone intense chemical weathering but little oxidation, a (B)/C horizon with considerable oxidation, or a C_w horizon (white horizon, pallid zone) that has been formed by heavy de-ironization in much the same way the white horizons of certain Rotlehm varieties are produced, that is, by the migration of the diffusible iron hydroxides to the crystallization centers above the pallid zone.

Laterization is found not only in Braunlehm or Rotlehm formations in situ but also in Braunlehm and Rotlehm sediments. Laterization is much favored in soil sediments because the condition of constant and homogeneous humidity without stagnant water tables is much more assured than in soil profiles in situ, and in addition because there exist great reservoirs of diffusible amorphous substances that are conductible throughout the extensive sediment layers. In such laterite profiles, a slag layer 20 m. (65 ft.) thick or more can be formed.

The so-called aging of buried sediments of very different origin, sometimes buried at great depths, may result in transformations similar to those of laterization. These are diagenetic transformations, produced not by the influence of the environmental conditions of the biosphere but by the crystallization of amorphous substances. Here also the matrix is such that a migration of diffusible peptized substances may occur.

Figure 12. Profiles of Laterization with Undisturbed Development in situ
from Braunlehm Sediment

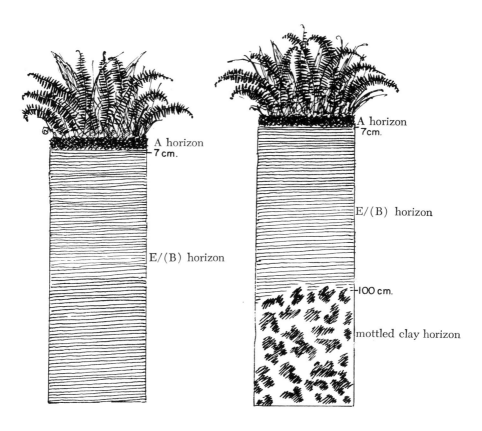

A. *Braunlehm Sediment*   B. *Lateritic Mottled Clay*

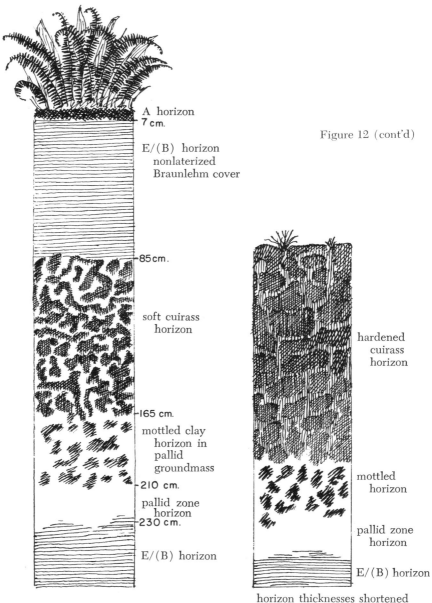

Figure 12 (cont'd)

A horizon
7 cm.

E/(B) horizon
nonlaterized
Braunlehm cover

85 cm.

soft cuirass
horizon

hardened
cuirass
horizon

165 cm.

mottled clay
horizon in
pallid
groundmass

210 cm.

pallid zone
horizon

mottled
horizon

230 cm.

E/(B) horizon

pallid zone
horizon

E/(B) horizon

horizon thicknesses shortened

C. *Soft Lateritic Cuirass*

D. *Hardened Lateritic Cuirass*
   after Erosion of the
   Braunlehm Cover

134

Figure 12 (cont'd)

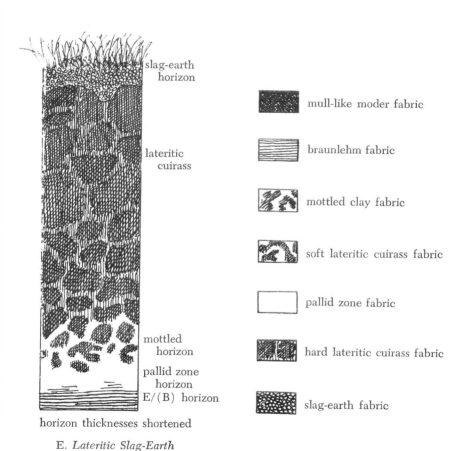

slag-earth horizon

lateritic cuirass

mottled horizon

pallid zone horizon

E/(B) horizon

horizon thicknesses shortened

E. *Lateritic Slag-Earth*

mull-like moder fabric

braunlehm fabric

mottled clay fabric

soft lateritic cuirass fabric

pallid zone fabric

hard lateritic cuirass fabric

slag-earth fabric

The foregoing description of a laterite profile applies to one which represents the final phase of laterization. Laterization proceeds in a number of different development phases, and although these phases are not clearly separated and may sometimes intergrade with each other, they influence profile formation considerably. Some profiles indicate complex development stages. Where laterite formation is a recent process (as in some parts of Río Muni), the whole development sequence of laterite profiles can sometimes be found within a small area, including the hardened cuirass exposed to the surface by erosion of the nonlaterized Braunlehm (or Rotlehm) cover. In Figure 12 several profile stages are presented, beginning with the formation of a mottled zone, continuing to soft cuirass layers, and ending finally with the hard cuirass exposed to the air and to desiccation.

As the final phase of laterization is reached, a new development phase sets in, and this one does not interest the pedologist as much as it does the prospector looking for iron and aluminum ores. This phase is characterized by a heavy leaching of the laterite layers. Old laterite cuirasses are, as a rule, highly permeable. Springs and creeks in old laterite landscapes therefore have a tendency to dry up, and, in general, one may observe other phenomena that are typical for a kind of karst formation found in limestone areas of the temperate zone.

In spite of the fact that the constituents of old laterite cuirasses have little mobility, a considerable leaching occurs in the course of immense periods of time as they are repeatedly attacked by abundant percolation waters in the rainy seasons. This leads to an enrichment of iron and aluminum ores. In spite of the slow solubility of secondary and primary quartz, this material may finally be removed completely from the fabric, thereby enriching the crystallized iron and aluminum oxides.

Iron hydroxide is usually present in the form of goethite, sometimes lepidocrocite, or dehydrated in the form of hematite. Some laterites contain magnetite; aluminum hydroxide crystallizes and accumulates in the form of gibbsite, sometimes of böhmite.

MICROMORPHOLOGY OF BRAUNLEHM-LATERITES

Laterization is possible as a transformation process from a braunlehm or a rotlehm fabric. Since the braunlehm fabric is much more suitable for laterization, the micromorphology and micromorphogenesis of a Braunlehm-Laterite will be described first and then the differences between it and a Rotlehm-Laterite will be noted.

Phase I: Formation of concretion layers. The formation of amorphous iron hydroxide concretions is a common phenomenon of the braunlehm dynamics; however, Braunlehm layers abundant in these concretions display a marked tendency to laterization in Río Muni as well as in the tropics in general (but not in the subtropics and not in fossil Braunlehm sediments of the temperate zone). Therefore, the formation of these concretions need not be fully discussed as a phase of laterite formation. Generally, there are sharp limits between the concretion horizons and the surface cover of unaltered Braunlehm. The concretions may be seen as formations in situ; or they are easily recognizable as accumulated erosion deposits.

Phase II: Mottling and encasing. In this phase, stains of a reddish-brown (2.5 YR 3/4–5 YR 6/8) or, in vertical light, bright-red (2.5 YR 6/8–10 R 5/8) color are visible in the still dense, yellow matrix of the thin section (Plate 12A). They are translucent, with a loose, flaky inner structure, and their delimitations are diffuse and irregular. At all marginal surfaces of the peptized matrix, that is, where there are fissures or root channels with embedded mineral grains, there forms a thin crust of precipitated iron hydroxide of a dark-brown (7.5 YR 3/2) or, in vertical light, bright-red (7.5 R 6/8) color. Mineral grains or fragments of the siallitic matrix that are exposed by fissures can thus become entirely encased in iron coatings. The same materials precipitate in cracks of the minerals and in small spaces in the interior of the matrix. The precipitates are produced by crystallization and consist mainly of goethite and hematite aggregates.

Phase III: Cuirass formation. As a result of aging and progressive dehydration of the gels and crystalloids, the precipitated masses contract and the former loose, soft iron sponge is transformed into a mass with a great variety of inner structures. The precipitated iron hydroxides darken and become almost opaque (Plate 12C and 12D). In vertical light they are bright red (7.5 R 4/8, partly 7.5 R 4/4). The amorphous braunlehm concretions crystallize and turn bright red. Like most of the other crystallized iron hydroxide masses, they are opaque and must be investigated with incident light. Since the peptized silica-rich matrix is interspersed by a firm filter skeleton, it can now be washed out easily, leaving a porous residue and empty spaces. Isolated matrix fragments, which are entirely encased in iron coatings, are protected from removal (Plate 12C). Since the silica in them also ages and is transformed into opal, they harden and their forms become stable when moistened. The coatings continue to build up as iron hydroxide migrates from the interior by diffusion. Because of this the matrix fragments become lighter in color and may be colorless at the end of this phase.

Phase IV: Interspersion with earthworm tubes. Earthworms are much attracted by the cavernous cuirass layer and deposit their excretions in it, for under moist conditions the cuirass has not entirely hardened. The casts consist entirely of unchanged Braunlehm material that has been derived from the fresh surface layer.

As a result of the intense earthworm activity, the reddish-brown cuirass layer is interspersed by numerous ocher or, finally, whitish stains. The filler materials can generally be recognized as earthworm casts by the naked eye. In some regions, earth termites instead of earthworms invade the cuirass layer, but never are the two found together.

The deposition of fresh Braunlehm material in spaces and tubes of old cuirass layers by earthworms or termites affords extensive possibilities for studying the genesis of the different phases of Laterite development since the genetic phenomena are repeated in the same order.

Phase V: Concretion formation in the Braunlehm tubes. Concretions are produced in the same way as in other braunlehm fabrics. The color of these concretions is also dark brown, but they never measure more than a few millimeters in diameter for lack of space and available mobile iron hydroxides.

Phase VI: Formation of vermicular structure elements. The Braunlehm tubes become encased in iron hydroxide rinds, as in phase II; the reddish mottling is less common.

Phase VII: Hardening of the Braunlehm tubes. Tube sections that are completely encased are protected from removal. With aging, the silica of the tube fillings is transformed into opal and, after further dehydration, into crystalline silica minerals, in which process the braunlehm structure hardens. At the same time, the remaining peptized iron hydroxide of the braunlehm matrix migrates into the surface rinds, leaving, in general, completely bleached tube fillings. Because of the mechanical breakdown into angular microscopic fragments, the whitish masses appear earthy to the naked eye. Since the silica is completely preserved, the total chemical composition of the tube fillings is no different from that of the original Braunlehm (Gallego, 1956).

During all phases, the transformation layers remain moist. The dehydration of the gels and crystalline iron and aluminum minerals is not produced by drying but by aging, that is, by contraction of the matter. On the other hand, in no case of genuine laterization could we observe stagnation of groundwater, reduction of the iron compounds or Rasen-eisenstein formation by reoxidation. In general, the soil mass during phase I and during part of phase II is conductible only by diffusion and not by gravitational eluviation. During the later phases, intensive eluviation is observed.

MORPHOLOGICAL VARIABILITY

The micromorphology of laterite formations is almost unimaginably varied both as to form and color in the different varieties of Laterite. Such variety is characteristic of laterization, whereas great uniformity is characteristic of rubification. A series of new laterite fabrics and fabric elements has been described by my former collaborator R. Schmidt-Lorenz (1964, 1967, 1970), who worked primarily in southern India in the laterite areas first described by Buchanan (1807) in Ceylon, and also by Hamilton (1964a, 1964b), who worked in central West Africa.

ROTLEHM-LATERITES

Laterization can also take place in rotlehm fabric provided no earthen-ing and leaching of the matrix have occurred. It depends primarily on the amount of active amorphous iron hydroxides left in the matrix. Rotlehm-Laterites differ from Braunlehm-Laterites in that there is a reddish matrix, which is only partly capable of becoming de-ironized, instead of the yellowish to whitish groundmass between the crystalliza-tion aggregates in the mottled phase. Also, in phase IV, since depositions in the spaces and tubes consist of Rotlehm material instead of Braunlehm fillings, these for the most part retain their red color. Hardened cuirasses of Braunlehm-Laterites are as a rule more compact than those of Rotlehm-Laterites.

SOIL FORMATION ON OLD LATERITE CUIRASSES

Laterites are among the most unfavorable parent materials for soil formation. Thus it is understandable why no effort has ever been made to investigate the so-called slag-earths, and why they do not appear in any system of soil classification. They are formed in areas in which the old cuirasses form surface layers as a result of a complete removal of the highly erodible Braunlehm or Rotlehm covers. Particularly in alternating humid and dry regions, a superficial breakdown of the fabric is gradually produced, giving rise to the formation of layers of lateritic slag-earth which have no plasticity. In a humid state they possess a water-stable fabric rich in spaces, but in prolonged dry periods they become pulverized and are easily eroded. They cannot be used for crop production and may serve at the most as rather poor pasture land.

CULTIVATION PROBLEMS IN LATERITE AREAS

Obstacles to crop production exist not only in the semidry areas but also in the ever-humid equatorial zone of the tropics. This is reflected in the fact that virgin forests left essentially in their original state are sparsely populated. Here, hunger and malnutrition are almost as common as in the subtropics and in the zone of alternating humid and dry tropics. The majority of the soils of the ever-humid equatorial zone are subject to rapid loss of fertility as a result of intense chemical weathering. Moreover, they suffer from the excessive humidity, often from dystrophism of their humus form and biology (Podzol region of the Amazon); and they have a dense and unstable structure and low permeability. Furthermore, numerous plant diseases flourish under the ever-humid conditions.

Clearing and cultivating soils in which the processes of laterization have set in present particular problems. Hardened laterite cuirasses are of course easily recognized, but more effort is needed to determine whether there is laterization *in statu nascendi*. We know that the precipitation of the amorphous sesquioxides in the form of coherent crystallization aggregates develops gradually, and that under the forest cover the process takes place without hardening of the laterized soil mass; however, as soon as the forest is cleared and the soil is tilled many hard clods, lumps, stones, crusts and rock fragments are produced, so that cultivation is almost impossible. Subsequently, a great many such soils that have been cultivated lose their earth content as a result of heavy erosion and leaching, leaving a stony surface. These soils support almost no vegetation and have become useless. It is therefore important that the tropical soils expert resort to thin section analysis to recognize the degree of laterization of a soil in its original state. In this way he can foresee possible unfavorable consequences of cultivation.

XIX

The Soils of the Tropics and the Edaphoids

Soil-like formations may be produced in nature without weathering and biological influence. Edaphoids, as the author terms these formations, are not soils, although they have often been mistaken for soils, being confused particularly with some tropical soils (Kubiëna, 1959). It is therefore of importance to know the micromorphology and mineralogy of these formations. They are produced mainly by postvolcanic action. As such, they may be formed either directly as volcanic products (soft pozzolanas, hydrated volcanic ash) or through the influence of volcanic heating and steam and gas exhalations.

Humid soils, as well as those of subaquatic, semiterrestric and terrestric nature, may also be excessively heated in volcanic areas, and thus may be transformed in a way completely unlike that which would have corresponded to their original environmental conditions. After an eruption, a volcano may return to seemingly complete dormancy, but high temperatures frequently prevail below the surface in some places for a long period, sometimes for centuries. The high temperatures may be sufficient for the formation of edaphoids. In addition to the above postvolcanic influences, local exhalations of hot gases may occur in the form of fumaroles and solfataras. These gases—hydrochloric acid, sulfurous oxide, sulfurous acid, sulfureted hydrogen, carbonic acid—are exhaled from volcanic magmas. They exert great influence on the transformation of soil minerals and rocks. This is chemical decomposition but not weathering or soil formation.

All these phenomena appeared in a past before there was life, before the environmental conditions of a biosphere and the development of true soils were possible. Edaphoids are therefore probably older than soils, that is, they were produced in periods in which no true terrestric soils, not even primitive raw desert soils (yermas), could have been formed. This has special importance for paleopedology because many edaphoids bear great similarity to a number of tropical soils. In volcanic regions, edaphoids with fabrics similar to those of tropical Braunlehm,

141

Rotlehm, Roterde, Pseudogley, Laterite and lateritic transformation varieties can be produced (Plate 13).

This is easy to understand, because humid tropical soils are formed in consequence not only of constant or alternating humidity but also of constant high temperatures. These conditions simulate volcanothermic influences, which are even more intense because of the effect of much higher temperatures than could be produced by a tropical climate. Not only can products similar to tropical soils be formed, but these products can be formed much more quickly and in a greater variety of environments. In terms of composition, volcanothermic action produces a number of minerals which cannot be formed in the corresponding soil types.

Fabrics similar to earthy Braunlehm, Braunerde, Brown Yerma and alpine or polar Braunerde have not been observed in edaphoids. Whereas one or another of the different edaphoid varieties may cover a sizable area (particularly braunlehm-, rotlehm- and laterite-like forms), the variety may change considerably within a few feet. This latter fact makes it easy to distinguish the edaphoids from true soils at first sight. Edaphoids may occur also in the form of sediments.

For micromorphological investigations, it is well to remember that edaphoids can become a parent material for soil formation in almost any kind of environment anywhere on earth.

BRAUNLEHM-LIKE EDAPHOIDS

The braunlehm-like fabric of edaphoids may be very pure and homogeneous, consisting only of a vivid egg-yellow matrix which does not contain any mineral grains of silt or sand size (Plate 13A). It shrinks considerably as it dries, leaving wide cracks between the joints. In such a matrix only colloids may be present; even the clay substance may consist of allophanes. Such material has been repeatedly described and investigated as a parent material for soil formation in southern Chile (Mella Lagos, 1958), Central Africa, New Zealand and Australia. Brown earthening produces an earthy Braunlehm without changing the amorphous clay. Rubification and red earthening give rise to its transformation into kaolinite (Besoaín and García Vicente, 1962). The presence of amorphous clay substances in highly decomposed edaphoids does not mean that edaphoids never contain crystalline clay minerals. Well-developed montmorillonite, for example, is common in edaphoids formed on basic rock (Mückenhausen, 1967).

Braunlehm-like edaphoids occur frequently in the form of filling substances (true bolus) in the vesicular spaces of volcanic lava. These bolus masses are amorphous, show no double refraction under crossed Nicols, and are soft and flowing, so that they move easily into cracks of the scoriaceous lava. The color of all these braunlehm-like edaphoids is much more vivid than that of true braunlehm horizons produced by weathering. Very striking colors, different from those of the corresponding soil formation, are also found in other varieties of edaphoids, such as the rotlehm- pseudogley- and laterite-like edaphoids. The coloration of the edaphoids therefore helps to distinguish them from soils.

BRAUNLEHM-LIKE FABRICS IN SILICEOUS POZZOLANAS

Pozzolanas are scoria particles which consist largely of volcanic glass and which, because of their large content of gases at the time of their formation, are interspersed with a large number of vesicular spaces. These pozzolanas not only may contain soft, braunlehm-like bolus masses in their pores but the solid parts of the vitreous rock may display a perfect braunlehm-like fabric of a vivid egg-yellow color. The ferric hydroxide is completely peptized and amorphous and exists as an extraordinary homogeneous substance within the silica groundmass. This shows that the siliceous groundmass can act as a peptizer for ferric hydroxide in the pozzolanas as well as in the soft braunlehm-like bolus.

THE BRAUNLEHM-LIKE EDAPHOIDS OF THE TRUMAO REGION, CHILE

The edaphoids of central Chile have a very clean and homogeneous braunlehm fabric consisting only of a very dense matrix which, under the microscope, looks no different from a braunlehm matrix except that its color is much more glaring and unnatural. In addition, they have no interspersed mineral grains, whereas Braunlehm soils almost always do. These edaphoids have been studied by my former Chilean collaborator, A. Mella Lagos (1958).

Wind deposits of these materials are widely distributed in central Chile and have become the parent material of different soils, notably of the Trumao series, which has the micromorphology of an earthy Braunlehm (Plate 13B). Some wind-transported sediments are completely rounded. In the process of drying, their fabric becomes highly stabilized, with the result that these aggregates are not destroyed by wetting. In spite of that, the secondary transformation of this volcanothermic material to earthy

Braunlehm takes place. All constituents of the matrix of the unchanged edaphoid are amorphous, including the clay substance. Only in the secondary transformation are these allophane clays changed into crystalline varieties—such as kaolinite, if rubification or pseudogleyzation takes place. These transformations have been verified by Besoaín and García Vicente (1962).

NONEXISTENCE OF EDAPHOIDS RESEMBLING EARTHY BRAUNLEHM

Earthy Braunlehm formed from braunlehm-like edaphoids is also found in Australia and New Zealand, but not as an original edaphoid with a fabric resembling earthy Braunlehm. This fact is significant. Volcanogenic edaphoids are products formed by the influence of humidity and high temperature in much the same way as the soils of the humid and alternating humid and dry tropics, which soils they resemble. This explains why soils of the kind that develop under the influence of a seasonal or diurnal change of temperature, with cool to cold periods, or like conditions, are not found in the depth of decomposition layers formed by volcanothermic influence.

ROTLEHM-LIKE EDAPHOIDS

Reddish-yellow, orange or bright-red edaphoids are common. The micromorphology of their fabric seems to correspond to that of a Rotlehm, because there are varieties which are only slightly reddish, no single crystals being visible under the light microscope (as in Rotlehm they are so small that they can be recognized only under the electron microscope). In addition, there occur all the transition stages, with fabrics that resemble those of earthy Rotlehm varieties, according to the degree of earthening. In these stages the iron hydroxide precipitations become visible as they flocculate to larger aggregates of a bright-red color. This process resembles the red earthening in soils with Rotlehm character (Plate 13C, 13E and 13G).

We can assume that the fabric of the rotlehm-like edaphoids is formed under conditions which cause rapid crystallization of the amorphous iron hydroxide. In some other varieties, in which slow crystallization has taken place, one sees dense crystal aggregates in a bright-yellow, later whitish, de-ironized matrix, instead of tiny single crystals.

The rotlehm-like edaphoids can be distinguished from Rotlehm soils by their dazzling, unnatural colors. Like all edaphoids, they will display marked variations in mineral decomposition as well. No data are yet available on this subject.

PSEUDOGLEY-LIKE EDAPHOIDS

Pseudogley-like fabric is particularly common in edaphoids (Plate 13D). With comparative investigations, it is easy to reconstruct the complete genesis. It begins with a humid braunlehm-like material in which bright-red mottles of iron hydroxide are produced by crystallization of the amorphous and peptized iron hydroxide of the matrix. The iron is moved by diffusion towards the crystallization centers. This process may continue until all the amorphous iron hydroxide in the matrix is used up. It is accompanied by a change of color in the matrix, which turns light yellow, grayish, and even completely white with complete de-ironization. Instead of mottling, the crystallized iron hydroxide may precipitate on the pore space walls and in cracks or in rounded vesicular spaces produced by gas inclusions.

Pseudogley-like fabrics are produced not only from soft, clay-like edaphoids but also from other materials, even solid volcanic glass. For the last, diffusible amorphous iron hydroxide must be present, since it diffuses easily through the solid silica matrix. For pseudogley-like edaphoids to form, no temporary stagnation of water is required, as is the case with Pseudogley soils. It may be assumed, however, that a more or less constant moistening of the substrate is produced by ascending vapors.

LATERITE-LIKE EDAPHOIDS

Amorphous iron hydroxide can migrate to the crystallization aggregates to such an extent that they grow large enough to touch each other and finally invade all parts of the former braunlehm-like matrix (Plate 13H). Heavy ironstones that can be produced by this process have been repeatedly confused with laterite formations. In addition to stain-shaped crystallization complexes like those seen in laterites, crystalline precipitations may appear in cracks and small crystal aggregates may grow laterally and join in honeycomb-like formations in laterite-like edaphoids. As in true Laterites, residues of the braunlehm matrix may be included between the iron precipitations; such residues may still contain some yellow amorphous iron hydroxide or be completely de-ironized to a grayish or whitish color. Like most of the edaphoids in volcanic environments, the laterite-like varieties are much brighter and more dazzling in color than their soil counterparts, the true Laterites.

HYDROTHERMAL GRAY MUDS

Hydrothermal gray muds are postvolcanic formations that originate from hydrothermal activity. They are characterized *in statu nascendi* by a high water content (hence, thermal muds). As they dry they shrink like many subaqueous and semiterrestric soils. Their color is affected by the iron content of the parent rock, and as long as no oxidation takes place as a result of contact with the atmosphere, they vary from intense dark blue-gray to very light gray. Contact with the air makes them change to brown, yellow or bright red.

To the naked eye, the gray hydrothermal muds look very much like Gley soils. Their sediments also have almost the same outer appearance as Gley sediments. In thin section analysis, precipitations of ferric hydroxides are rarely seen; on the other hand, precipitations of iron sulfide are very typical, just as they are in the Gley layers of marine warp soils. Nonetheless, these volcanic gray muds cannot otherwise be likened to Gley soils. Their most characteristic feature is not the presence of reduction, although the influence of many gases exhaled by fumaroles may be strongly reductive. However, gray clay muds may also be formed in hot springs where there are no gas exhalations. The most characteristic feature of the hydrothermal gray clays is lack of oxidation (although oxidation can readily be produced). In this respect they are similar to the C_m horizons of extremely humid soils formed on easily weathered basic rocks in the tropics.

The author found the most typical volcanic muds in Yellowstone National Park, in Tenerife, in the Solfatara Pozzuoli, near Naples, and finally in Iceland. Knowledge of the Icelandic gray muds has been of special importance to him in view of the thorough mineralogical and chemical investigations of these formations by Gudmundur E. Sigvaldason. His descriptions are fundamental not only for the Icelandic formations but for hydrothermal gray clays in general. The following partial characterization of some typical varieties is based on his investigations.

GRAY MUD OF FUMAROLES FROM BASIC ROCKS

This formation is produced by the disintegration of the rock under the influence of acid solutions, primarily by a high concentration of sulfuric acid in conjunction with high temperatures.

For his investigations of this mud, Sigvaldason used a variety derived from basalt north of Hveragerdi in the southwest of Iceland. The author had occasion to see the habitats of these formations and to take samples for the preparation of thin sections. Sigvaldason had seven samples representing increasing degrees of decomposition, beginning with an almost undecomposed basalt and ending with the boiling clay mud at the opening of the fumarole (sample 7). Here the mud had a pH of 1.5. In samples 6, 5 and 4, that is, at increasing distances from the fumarole, the values changed from pH 2.0 to pH 3.5. The considerably decomposed matrix was not peptized but had a marked aggregate formation which was clearly visible in the thin section.

At the beginning of the decomposition, new formations of calcite and montmorillonite were produced. The accumulation of montmorillonite reached its maximum in samples 2 and 3, but decreased considerably with continued alteration and was replaced by kaolinite. With increasing decomposition, great quantities of free colloidal silica (like that in humid tropical soils) were produced, and these soon accumulated in the form of opal (opalization in tropical soils is a much slower process but leads finally to the formation of secondary quartz). In addition to iron sulfide, we found hematite and goethite (needle ore), primarily in parts of the fabric that had been exposed to the oxygen of the air (red seams on the edges of cracks in the drying mud, or sections of red clay formation).

In the thin section of a gray mud sediment dug open at the Sanatorium of Hveragerdi, I found a secondary lessivé fabric. The gray matrix, consisting of aggregates matted together into a coherent mass, was interlaced by secondary conducting grooves filled out by an ocherous clay plasma (stained by peptized iron hydroxide). The fluidal plasma had a higher double refraction produced by particle arrangement. The partly dried mud, which evidently had suffered repeated redeposition (since it consisted mostly of fabric fragments), displayed new open shrinkage cracks. Here, small octahedrons of pale pyrite crystals were visible under a vertical light microscope.

GRAY MUD OF FUMAROLES FROM ACID ROCKS

This mud is characterized first of all by its light-gray color, which results from its low iron content. In all decomposition phases, kaolinite is the only clay mineral; it is always represented by well-crystallized forms.

Sigvaldason analyzed a gray mud of fumaroles around the glacier Torfajökull in southern Iceland that had formed from liparitic rocks. Like the decomposition of basalt, the alteration was accompanied by an increasing extraction of the alkalies and of calcium and magnesium, with the loss of sodium considerably higher than the loss of potassium. Calcium was completely dissolved, whereas magnesium proved to be more stable.

GRAY MUD OF ALKALINE HOT SPRINGS

The characteristic minerals for this variety are montmorillonite and zeolites. Both are very stable and are present also in the final products of decomposition.

Sigvaldason investigated a gray mud of the hot spring Spytir in the Hengill-Hveragerdi region which had originated from a dense, fine-grained basalt. The boiling spring water (100°C.) had a pH of 9.4. The clay mass was somewhat brownish and was covered with a whitish-gray sinter layer. The radius of efficiency of the alkaline solutions was considerably lower than that of the acid solutions. At a distance of 2 to 4 meters (6½ to 13 ft.) the basalt remained practically intact. By comparison with the acid solutions, the alkaline produced less intense decomposition. The montmorillonite content increased until the final phases of alteration, attaining values up to 80 per cent. In all samples, pyrite was a common mineral. It usually occurred within a material that could be identified as zeolites. The pyrite occurred almost always in cubical crystallization, octahedrons being rare. The fact that hematite was not found allows the conclusion that reducing conditions are much stronger in alkaline hot springs than in acid fumaroles and solfataras. Loss of elements with only a few cations could be observed (sodium, calcium and magnesium). There was a large accumulation of potassium, perhaps because of its absorption by montmorillonite.

XX

Soils of the Semiarid Regions of the Tropics and Subtropics

AC SOILS

Since the chemical weathering that takes place in the ever-humid tropics is the most intense on earth, the initial soil development stages, the AC soils, the Ranker and Rendzina varieties, are rare in such latitudes. By comparison, in the alternating humid and dry tropics, the development of the AC soils does take place to some extent, even though chemical weathering as a result of the very humid rainy seasons is still very intense. The Ranker formed under these conditions is almost always a Mull-Ranker.

A very typical variety of AC soil in the alternating humid and dry tropics is a soil known as Ganter. It occurs only on easily weathered parent rocks (such as basalt, gabbro, norite and melanocratic diorite) and is characterized by a gray C_m horizon. This horizon is produced by the intense chemical weathering, but without much oxidation. Like the parent material, the weathered decomposition mass has no cavities, and therefore air and oxygen cannot penetrate it easily. Ganters are Ranker-like mull-forming AC soils that are strongly weathered and contain clay accumulations that form a dense, clayish, fragipan-like gray C_m horizon between the A horizon and the parent rock. The mull of Ganter is easily transformed into tirsoid humus formations. The fabric becomes dense and anmoor-like in the A horizon, and this stabilizes the AC profile considerably.

Soils of this kind occur in India and Indonesia, and in this part of the world they are called margalite soils. This grouping, however, includes calcareous varieties that are not Ganters in our sense, but ACaC soils, which have affinities with the Rendzinas or, more often, the Smonitzas.

TIRS SOILS

Tirs soils are very characteristic of the alternating humid and dry tropics and subtropics and correspond to most of the swamp soils, Anmoors and Gleys of the humid tropics. A Tirs is formed on an imper-

149

meable substrate such as a C_m horizon, a Braunlehm or a Rotlehm sediment. This substrate causes the surface water to stagnate in the wet seasons, and the humus horizon to dry up in the dry seasons, and hence there is a seasonal change in the biology of the soil. The soil life can be compared with that of an Anmoor in the wet season or with that of a mull in the dry season. According to the character of the humus, it has features of both humus forms. It is gray to light gray in the dry state and almost black in the wet state. The structure is very unstable, and the fabric is dense and highly erodible. The humus horizons can attain great thickness, as much as 75 cm. or even a meter (30 to 40 inches). The lower part of the A horizon has usually undergone pseudogleyzation, particularly when formed on Braunlehm sediments.

Micromorphological investigation reveals, in addition to the dense structure and shrinking and irregular distribution of the humus, microfossils typical of a semiterrestric soil life, such as shells of diatoms, calcareous shell fragments and blackened xyloid plant splinters. These fossil forms are found in plains areas, in depressions and on lower slopes, but not on hilltops and upper slopes. They occur also in the tropical savanna, and in the subtropics or some parts of the mediterranean zone.

ENVIRONMENTAL CONDITIONS OF THE SUBTROPICS

In planetarian classifications of the climates, a subtropical zone sometimes is not considered since some investigators are inclined to regard it as a part of the mediterranean or even the temperate zone. The subtropics constitute a sunny zone with warm summers and mild winters, situated between 20° and 40° north or south latitude; the air pressure is high, and the sky is usually bright, rarely overcast. This zone, however, is not clearly delimited. In general, the areas in which palms become native trees are accepted as marking its northern limits (for example, where *Chamaerops humilis* grows in Spain, or the Washington palm *Washingtonia filifera* in southern California).

In soil geography it is necessary to distinguish a subtropical zone. Although the soils in this zone are similar to those of the ever-humid and alternating humid tropics, there are typical differences. The most important has to do with the distribution of Laterites; these soils are typical for the tropics but are extremely rare or nonexistent in the subtropics (the few that do occur are relict soils).

SOILS OF THE SUBTROPICS

Braunlehm, Rotlehm and Roterde occur in the subtropics in forms similar to those of the tropics. As to mineral content, they display increasing amounts of micas, illites and vermiculites. However, with detailed investigation one recognizes that these Braunlehm, Rotlehm and Roterde varieties in the subtropics have usually undergone considerable transformation.

Typically humid Braunlehm and Rotlehm varieties are rare. They are regressive and give way to soils whose behavior must be regarded as the behavior of typically dry soils. To these varieties belong the following: the xeromorphous Braunlehm and Rotlehm varieties, the Dry-Rankers and the various desert raw soils.

In thin section analyses, a transformation can be recognized in its first stages, and my comparative investigations show that the semiarid and arid soils are definitely progressive, in contrast to the regressive humid soils. This is true not only within the limits of the subtropics but far into the zone of the alternating humid and dry tropics and to some extent even into the formerly ever-humid tropics.

PROGRESSIVE XEROMORPHISM

The causes of progressive xeromorphism are primarily of a general climatic nature. In the search for relict soils, evidence was found that in former geological periods the humid tropical and subtropical soil varieties of Africa extended to the coast of the Mediterranean Sea and beyond. According to Büdel (1955), every glacial period in the Pleistocene age corresponded to a humid period and every interglacial to a dry period in the alternating humid and dry tropics and in the subtropics. This opinion is not yet generally accepted, but the case surely rests with results of paleo-pedological investigations. The border of the humid tropical soils moved northward in the humid periods, and their relicts were left behind in the dry periods. Although they were transformed by the new environmental conditions and to a great extent denuded by increased erosion, their residues constitute valuable evidence of the earlier humid environmental conditions.

It is very clear from thin section analysis that xeromorphous conditions like those of the present also prevailed in some previous periods, thereby interrupting the effects of humid climates. That these must have been periods corresponding to the interglacials in the north can be concluded from the fact that our present age does not correspond to a glacial but rather to a kind of interglacial period in the temperate latitudes of today, with practically the same soil formations. In all parts of the present tropics and subtropics, moreover, the rare relict soils of earlier humid climates are interspersed with the common soils of the semidry or dry environmental conditions of the present.

Besides the natural tendency toward increasingly dry climatic conditions in the present age, a very significant new factor must be considered. This is the influence of human activity. To this factor my colleagues and I have given special attention in our comparative investigations in different parts of Africa, the Cape Verde Islands, the Canary Islands, Madeira, the Azores, Ecuador, Brazil and other parts of South America, and Central America. The kind of use made of the land by man is of crucial importance to the future.

Where too much timber is cleared and where too many goats and sheep are allowed to graze in tropical and subtropical regions, the soils deteriorate rapidly. So destructive are the grazing habits of these particular animals that they not only prevent natural afforestation but destroy new forests that have been planted for conservation purposes and are in their first stages. In those countries in which there is intensive farming in spite of the decline in rainfall, soils that are cleared and cultivated and exposed to the hot sun in a succession of drought seasons quickly lose their soil life and humus content; the thickness of the humus horizon decreases, as does the intensity of chemical decomposition. With the deterioration of the soil structure, the soil decreases in fertility, turning finally into wasteland. Soil erosion sets in promptly and removes what is left of the once productive soil.

Our observations are fully supported by specialists in geobotany, forestry research and historical geography. It is therefore of the utmost importance that the many young states in the subtropics and the alternating humid and dry tropics be fully cognizant of the soil problems that must be considered in all planning for agriculture and forestry and the general development of the land.

XEROMORPHOUS TRANSFORMATION OF BRAUNLEHM

Humid Braunlehm is not a very fertile soil. Because of the intense weathering and excess of moisture, it is poor in nutrients, so that even in the humid tropics its fertility decreases rapidly and forces the native farmer to abandon it after a few years and clear a new patch of forest for cultivation. Other great disadvantages of humid Braunlehm are its extremely dense fabric, high erodibility and lack of stability, lack of aeration, low permeability, and, therefore, its poor soil life and humus formation.

If, by a drastic change of the climatic conditions, a humid Braunlehm is exposed to a semidry climate, the unfavorable properties of this humid soil are considerably worsened. The dense structure of the humid soil is associated with a soft soil mass. In a dry climate, a very troublesome crust formation in the surface layer may be produced.

The great importance of soil structure for crop production is at last fully understood. A soil may be well provided with nutrients, showing not the slightest deficiency in this respect; but if it has the structure of one of the solonetzic varieties, which do not support higher plants of any kind, a high nutrient content will not change its value. The crop yield will still be zero.

The case is similar with the transformation varieties of Braunlehm in semidry regions. With intensive farming, the crop diminishes each year, until finally the soil is of no use even as inferior pasture land. Some drought-resistant weeds grow in patches, but most of the soil surface remains bare. From the Spanish *calvo*, bare, the transformed soil is designated as Calvero-Braunlehm. This soil variety is distributed over large areas in the semidry subtropics and the alternating dry and humid tropics, but for the most part has been abandoned by man.

The recovery of the soils in these areas for agriculture and forestry is an important problem in practical soil science. It can be resolved only on the basis of a thorough knowledge of the nature of these soils and of their development tendencies.

Arturo Primavesi (1961) of the Federal University of Santa Maria, Brazil, has conducted some very effective field studies on the problem in the alternating humid and dry zone of Brazil. Primavesi's experiments are in a vast area in the state of Minas Gerais near Salto the Pirapóra at 21° south latitude, where the Calvero-Braunlehm soil had changed into near-desert, with heavy compaction and a groundwater level at 38 m. (125 ft.). *Aristida pallens,* growing in patches between the barren tracts, and some brushwood of *Xylopodium* species were the only vegetation. The region had not been cultivated for eighty-seven years.

Primavesi has concentrated on improving the soil structure, promoting a well-balanced soil life and correcting mineral deficiencies. Four years of biophysical treatment, with constant control of the biology, structure, humus formation, calcium carbonate content, pH and water economy of the soil, have finally made it possible to raise an exceptional wheat crop. An important part of the treatment is green manuring, mixing legumes into the very surface layer of the soil so as to form almost a mulch. Some check plots received the same fertilizers but no biophysical treatment. In 1958, an extremely dry year, the yield on the plots that had received full treatment was 1,806 kg. of wheat per hectare, whereas the wheat on the check plots died before heading because the soil had been exposed directly to the intense radiation of the sun.

The successful experiments of Primavesi thus demonstrate that it is possible to rehabilitate biologically, structurally and chemically degenerated Calvero-Braunlehms. Such soils cannot be restored by the mechanical and chemical methods commonly used in the humid temperate zones; the techniques must be adapted to the specific environmental conditions to which these xeromorphous Braunlehm varieties are subject. The rehabilitation of soils thus requires a highly specialized approach, and the techniques will vary according to the peculiarities of the climate as well as of the soils themselves.

PROGRESSIVE XEROMORPHISM IN THE EAST ATLANTIC ISLANDS

In the course of comparative studies of several islands of the East Atlantic, it was again apparent that the damage to soils from progressive xeromorphism occurs primarily in the subtropics and in the tropics with alternating humidity. The islands on which soil transformation was studied were the following: Santa Maria and São Miguel in the Azores (37° N), Madeira (32° N), Grand Canary, Tenerife and Palma in the Canary Islands (28° to 29° N), La Sal and São Tiago in the Cape Verde Islands (15° to 17° N) and Fernando Póo (3° N) and Annobón (1° S) off the Guinea coast. These islands are good subjects for comparative studies since most of their soils are produced on the same parent rock, a basalt that tends to yield aegirine-augites and olivines, and other minerals.

The recent soils in the Azores and Madeira are Mull-Ranker, humid eutrophic Braunerde and some Pseudogley (with many relicts of Rotlehm). In the Canary Islands (Figures 13 and 16), Xero-Ranker and subtropical Dry-Braunerde occur in the foothill belt, earthy Braunlehm in the montane belt (with many Braunlehm and Rotlehm relicts), and

Figure 13. Recent Soil Development on the Islands of Tenerife and
Fernando Póo

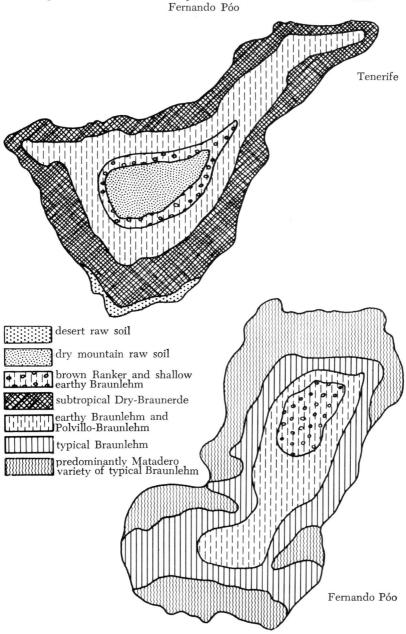

Tenerife

:::::::: desert raw soil

dry mountain raw soil

brown Ranker and shallow
earthy Braunlehm

subtropical Dry-Braunerde

earthy Braunlehm and
Polvillo-Braunlehm

typical Braunlehm

predominantly Matadero
variety of typical Braunlehm

Fernando Póo

Ranker and raw soils in the dry alpine belts. In the Cape Verde Islands, Rotlehm and Ranker to desert raw soils are found. In Fernando Póo, tropical Braunlehm occurs in the foothill belt and earthy Braunlehm in the montane belt. In Annobón, where the parent materials are very basic and iron-rich basanites, the soil is ferruginous earthy Braunlehm.

The most interesting observations were made in the Cape Verde Islands, where xeromorphism consists in a progression from Braunlehm and Rotlehm to desert raw soil as a result not only of climatic change but of the use made of the land by man.

SOIL XEROMORPHISM IN SÃO TIAGO AND LA SAL

The original vegetation in the lowland and foothill belt of these Cape Verde islands was savanna forest or forest savanna. Judging from the residues of the former soil cover, the Rotlehm extended up to an altitude of about 300 m. (1,000 ft.). Today neither Rotlehm nor the humid earthy Braunlehm of the montane belt is formed, although the latter can still be found in the montane belt of Tenerife and Palma. These soils have been transformed into powdery xeromorphous varieties. The forest vegetation has become very rare even in the comparatively less arid island of São Tiago; this island does have some remnants at altitudes above 1,500 m. (5,000 ft.).

The lack of a closed vegetation cover has favored and hastened the processes of degeneration and pulverization of the soils. They suffer not only from intense desiccation under the hot sun in the dry season, which lasts from January until July, but also from the assaults of the strong northeast tradewinds. The valleys have lost a great deal of their fertility. There are no longer very many tropical plantations of coconut palms, bananas and sugar cane, and those that do operate are in very poor condition. Without irrigation, agriculture in these islands is always at the mercy of drought as well as the locust plague.

I happened to visit the island of São Tiago in 1960, a year of catastrophic dryness. The Portuguese governor in Praia, the capital, organized a large road construction project to enable the people to earn money to buy food, all of which is imported from the mainland.

Soil xeromorphism was most advanced in the island of La Sal. Here, also, Rotlehm was the original soil cover, but today Rotlehm is found only as a fossil soil. It has been covered by great deposits of volcanic ash, and in spite of the fact that volcanic ash weathers easily in the tropics and gives rise to the formation of thick soil covers, the present development

stopped with the phase of a desert raw soil. Sporadic relics or fossil profiles of Xero-Ranker are found only in isolated areas. The plant cover has
disappeared almost entirely, and La Sal has turned into a desert.

SOIL XEROMORPHISM IN THE CANARY ISLANDS

Soils in the Canary Islands have suffered much the same fate. Originally,
the forest extended all the way down to the shore line, at all points. On
the island of Tenerife today there is laurel forest in the north only at
altitudes between 400 and 1,800 m. (1,300 and 5,900 ft.), and in the south
only between 800 and 2,000 m. (2,620 and 6,560 ft.). Moreover, in consequence of the rapid xeromorphism of the vegetation from the coast
upward as well as from the dry summit area downward, the belt of laurel
forest is comparatively narrow, and even in this belt it grows only in the
favorable habitats. Earthy Braunlehm is still the soil cover in these laurel
forests, and this is typical of the fog woods of the humid tropics. A peculiar variety of that soil, Polvillo-Braunlehm, was found repeatedly. It consists of rounded, very stable aggregates, and usually is formed only from
basic parent rock (basalt) in the montane belt of the ever-humid tropics.

However, these earthy Braunlehm varieties of the Canary Islands
exhibit clear symptoms of degeneration. The profile is much reduced in
depth, and the humus horizon is much less thick. Decomposition and
humification are less intense, and there is a hardening of the polvillo
aggregates. The humidity of the soil is decreasing, and this is apparent
from the condition of the vegetation.

The dry soils of the islands are becoming drier very rapidly. In the
lowland areas the former Braunlehm varieties are replaced by a subtropical Dry-Braunerde, a Xero-Ranker and lime crust raw soils. In the
southwest section of Tenerife there are only desert raw soils, because the
present environmental conditions do not permit the development of a
Xero-Ranker or a Dry-Braunerde.

A sediment of small, rounded fragments of pumice stone (called *jable*)
that is present in this region is of great importance to the local mode of
dry farming. The pumice stone, with its excellent capillary system, permits
optimal reception and retention of moisture; its capillary system is in fact
superior to that of any soil. The fields of *jable* in the Canaries are arranged
in terraces, separated by stone walls, and are used mainly to grow
tobacco. Micromorphological and micromorphometrical investigations of
the capillary system of the *jables* of the Canary Islands have been carried
out by Kress-Voltz (1964, 1967).

XXI

Soils of the Arid Regions of the Tropics and Subtropics

ENVIRONMENTAL CONDITIONS OF THE DRY DESERT

Dry deserts are situated primarily in the high pressure zones between 20° and 30° north or south latitude, in regions where the evaporation exceeds the precipitation. Low cloud formation, high diurnal temperatures and strong nocturnal radiation are characteristic of these deserts. The vegetation, if any, is poor. The hot deserts of the Eastern Hemisphere grade into etesian climates to the north and into savannas to the south, towards the Equator. Other deserts, such as the deserts of Central Asia, are produced by the extreme continentality, their situation in the interior of large continents. Here the summers are hot and the winters very cold.

Because of the intense heating of the rock surfaces in daytime and the great temperature drops during the nights, air circulation is very strong, producing many high winds, whirlwinds and dust spouts. In consequence of the heavy wind action, the lack of protecting vegetation and the intense physical breakdown of soil and rock fabrics, deflation of sediments and soils is a typical phenomenon of the dry deserts. Erosion by flowing water is also severe, for the rare rains come suddenly and in the form of violent downpours that can generate enormous mud and stone floods. Typical for the topography are basins without exterior drainage. They collect not only sediments of different kinds but also large quantities of water-soluble salts. This may bring about the formation of salt lakes, and, as these dry out, they become clay plains with superficial salt efflorescences or hard salt pavements.

Life in the dry desert is possible only because of special adaptations in or by the organism. It is essential to conserve water. Some animals are endowed with protective arrangements against evaporation, such as thick body covers, and the ability to burrow and seek refuge in cooler ground. Man gains protection against evaporation and dehydration by completely covering himself with clothing. Whereas the natives in the equatorial zone originally did not wear clothes, the Arabs in the much hotter Sahara desert have always chosen to wear white garments down to their feet and to cover their heads with turbans. During the day, when outside their thick-walled huts, the Arabs can wind their turbans around their faces, leaving only the eyes free. Similar adaptations permit the bushmen of the Kalahari Desert to run 55 kilometers in five hours in glowing heat on loose sand, and their women to carry loads equal to their own weight for 5 kilometers through the scorching desert. But the life expectancy of the bushman is not great; he rarely lives to be more than twenty-five years old (according to a report by G. Silberbauer).

As to the consumption of fluids, I remember that on our excursions through the mountains of Ougarta, Algeria, and the Great Western Erg of the Sahara we were each allotted nine liters of water a day. The water was carried in a black goatskin to keep it cool, and although this gave it a yellowish color and an unusual taste, not one of us failed to drink all nine liters.

DESTRUCTION OF SOIL

With a few exceptions, the deserts of the earth are not original landscapes like those on the moon; they are scenes of old devastation and destruction. The paleopedologist invariably finds evidence that they once possessed humid soils with a rich vegetation and animal life, and even environments favorable to man (as, for example, in La Sal Island of the Cape Verdes). Numerous stone carvings in different parts of the Sahara, the work of prehistoric peoples, show hunting scenes with large numbers of gazelles.

Paleopedological investigations add enormously to our knowledge of past eras. The contrast between the true desert soils and the relict soils of former humid environments is much more striking in the arid zone than in the temperate zone. On the other hand, relict soils in the desert are much more subject to complete destruction and removal, so that only a few remnants persist. Under arid environmental conditions, transformation effects an almost complete breakdown of the soil fabric into highly dispersed formations which are subject to pulverization and wind erosion. Most of the pulverized, finely dispersed soil constituents are removed from the desert, and there remain only the coarse elements, the former fabric skeleton (the coarse sand and a part of the fine sand). The fine constituents are picked up and carried great distances by air currents, and ultimately they are deposited as thin dust layers in cities (trade dust in North Africa and Europe) or on the surface of snow covers in the Alps (producing blood snow in the case of Rotlehm dust) or in the high mountains of Greece and Crete; or the pulverized material may be brought down in the rain (blood rain, mud rain, miracle rain).

These deposits do not lend themselves to the formation of sediment layers. The pulverization processes in humid soils of the past in desert areas are easily studied by fabric analysis and by direct observation. Particularly well known are the red dust deposits in North Africa and Europe, in West Africa, Turkestan, China and different parts of South America. Investigations of a great dust deposit in Palermo showed that the average grain size was 0.012 mm., whereas the grain size in a dust deposit carried from the African deserts to Hamburg was some 0.009 to 0.014 mm. The waters of the Atlantic Ocean between Madeira, the Cape Verde Islands and the west coast of Africa, where the prevailing winds are often east winds, contain desert dust and are known as the Dark Sea.

Not all the relicts of humid soils are removed in this way. Particularly in the Sahara, there are many habitats in which the Rotlehm residues are well preserved. For one thing, old (B)/C horizons have remained in place without destruction because the pseudomorphism of their rock fabric was protective and made them resistant to erosion. On the basis of their thickness and kind of micromorphology, and with the aid of comparative investigations of recent soils, the investigator can usually reconstruct the former soil profile. Of the Rotlehm relicts that I have observed in my travels in the Sahara of Algeria, one variety, the white horizon variety, is the most common. Braunlehm relicts were found only in mountain areas and as sediments in the humid valleys of the oases.

Recent terrestrial soil development in the desert is limited to the formation of Yermas, the raw desert soils. These raw soils are poor in binding substances, and therefore display no aggregate formation; they are loose and open to erosion. An accumulation of this material is possible only in very protected areas, and here soil accumulations may have some soil life but it is not possible for a humus horizon to develop. The profile scheme is at best $(A)C_1C_2$ on solid parent rocks. Little chemical weathering takes place, almost no clay accumulates and there is no earth formation at all. If enough chemical weathering takes place to give the material some color, it turns a weak light brown, much like mediterranean Dry-Braunerde. This type of raw soil is called a brown Yerma.

Red is a color that is never seen in these recent soil formations in the desert, though it is characteristic of relict soils and of some desert varnish. Chemical weathering and the accumulation of iron hydroxide take place too slowly to permit the formation of reddish soils in deserts. For this, at least the very wet rainy seasons of the savannas are needed. Even in the areas with Calvero-Braunlehm and mediterranean Dry-Braunerde, which have much more precipitation than the desert, bright-red soils of recent origin do not exist.

In the desert, free iron hydroxides accumulate in a striking way as desert varnish on the surfaces of boulders and rock fragments that are exposed to the air and to the radiation of the sun. No varnish forms on the underside of the rock. Occasionally the desert varnish is red, but generally it looks very dark, almost black. It is precipitated in a very dense, opaque form.

TRANSFORMATION OF RELICTS OF HUMID SOILS

Because little leaching takes place in deserts, the salt content of desert soils is usually relatively high. In addition, salts are easily transported by wind action from the salt-enriched surfaces of desert salt soils (mostly semiterrestric formations) or as salt sprays from the coastal areas to salt-deficient areas or to those poor in soluble salts.

The high solubility of these salts and the ease with which they precipitate by recrystallization in soil layers have an important effect on the coherence of the soil fabric. Using the same kind of salt in the laboratory, it is easy to produce either a total breakdown and destruction of the dense soil fabric or a complete cementation of the soil mass (such as has occurred in the *costra* formation in the desert of northern Chile). The result depends on whether one adds the salt solution in small quantities so that the fabric absorbs and crystallizes it continuously, little by little, or whether one saturates the soil, producing total stagnation of the liquid, then allowing it to concentrate slowly and the salt to crystallize into a coherent block. Both processes, the fabric destruction and pulverization and the cementation, occur in different habitats of the desert, the second being typical for semiterrestric soil formations. Very small amounts of salt may be enough to produce pulverization.

However, another form of fabric destruction that may occur has the same effect of pulverizing relicts of tropical humid soils in the desert climate. The binding substances of many of these soils have a high swelling power and shrink considerably with drying. With the extreme desiccation and heating that take place under the blazing desert sun, the binding substances, which unite the skeleton elements of the fabric to a coherent body, not only shrink to the utmost but lose their swelling power and binding capacity. With every new wetting and drying, the mechanical decay is increased, and ultimately the finely dispersed soil constituents are completely pulverized. Pulverization is very great in highly weathered, clayey, non-earthened Rotlehm and Braunlehm relicts (Plate 9D, 9E and 9F).

HYPSOMETRIC ZONATION OF SOIL DEVELOPMENT IN THE DESERT

With increasing altitude in the arid desert, the belt of recent raw soils (Yermas) extends (without intermediate belts) into the area of the mountain raw soils (alpine Råmark) of the nival belt, where environmental conditions are those of the cold desert.

With the increasing altitude, however, a hypsometric zonation can be seen in the degree of preservation and mode of transformation of the relicts of humid tropical soils. A good example is the transformation of relict soil with changing altitude in the Atakor of the Ahaggar, a vast high mountain plateau in the tropical part of the central Sahara (see Figure 14). The distribution of different relict and fossil residues in the different mountain belts is of interest in this connection (Kubiëna, 1955b).

PALEOSOILS IN THE ATAKOR

In the Atakor of the Ahaggar the well-preserved soil cover consists primarily of relicts of earthy Braunlehm. This is surprising because this soil is essentially a forest soil. Such soil also occurs in the montane belt of the Canary Islands, which supports the laurel and *Pinus canariensis* woods of the West Canaries; it occurs even in the montane belt of the equatorial zone, in the fog woods of Fernando Póo.

The degree of preservation of the Braunlehm relicts in the Atakor is surprisingly good, especially in view of the absence of forest cover. In the valleys there exist, besides raw desert soils, greatly broken down relicts of Rotlehm; Rotlehms are also found in the montane belt of the Atakor. These Rotlehms occur, not on the present soil surface, but as fossil formations between basalt covers of different ages; these fossil layers are in an excellent state of preservation.

OROGRAPHY AND VOLCANISM

The Atakor is a volcanic mass in the center of mountain country, the Ahaggar range, whose highest peaks attain an altitude of 3,000 m. (9,840 ft.), with many exceeding 2,500 m. (8,200 ft.). The Assekrem has an altitude of 2,720 m. (8,922 ft.). The fantastic landscape of the Atakor is composed of a series of basalt covers from which protrude an unusually large number of vertical rock needles, columns and steep domes of phonolite, trachyte and trachyandesite up to a relative height of 400 m. (1,312 ft.). In the central part of the Atakor an 800 sq. km. (500 sq. mi.) area contains three hundred extinct volcanic chimneys. There are, besides, deposits from old eruptions of different ages and extensive basalt covers, which make these mountains an ideal research area for paleopedological investigations, except for the fact that field trips have been difficult to arrange.

Fortunately, I was able to take advantage of the excursions organized by the Nineteenth International Geological Congress in Algiers in 1952. Trucks and riding camels were provided in our center in Tamanrasset, and in the course of our field studies I gathered soil samples for thin section analysis.

ENVIRONMENTAL CONDITIONS OF THE ATAKOR

Because of its altitude, the Atakor of the Ahaggar is much more humid than the rest of the Sahara. It is even more favored climatically than the other parts of the Ahaggar. The winter is distinctly cold with frequent frost, but the summits are snowcapped for only a short time. Occasionally there are heavy hailstorms. The spring is fresh and agreeable, the summer hot and dry. Because of these environmental conditions the Atakor is a kind of refuge for man, animals and plants. The vegetation, like the soils, has relict character. In the mountain belts one sees inferior examples of Mediterranean species; in the lower belts, Sudanese tropical forms.

BRAUNLEHM RELICTS IN THE ATAKOR

Braunlehm relicts are distributed throughout the Atakor of the Ahaggar, indicating that the soil formation of the past was generally a well-weathered, fairly thick soil cover. The parent material was, in the first instance, the older basalts; these occur in extensive lava covers and not in the more localized lava flows of middle-aged and young basalts. The acid lavas are present mostly in the form of rock needles, steep columns and bare domes, which, topographically, are ill-suited for soil formation. Braunlehm exists not only in profiles in situ but also in the form of deep erosion sediments that cover pediments and hammadas.

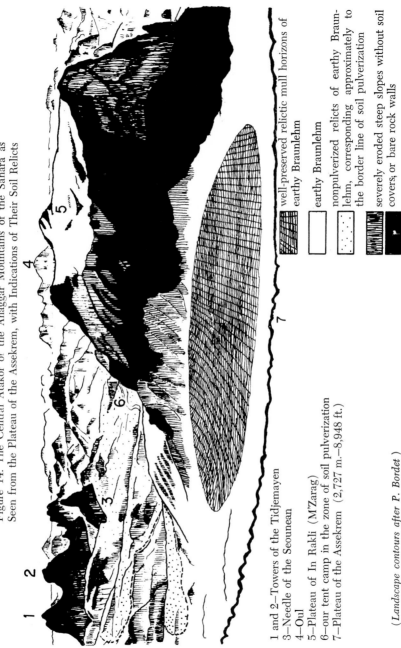

Figure 14. The Central Atakor of the Ahaggar Mountains of the Sahara as Seen from the Plateau of the Assekrem, with Indications of Their Soil Relicts

1 and 2–Towers of the Tidjemayen
3–Needle of the Seounean
4–Oul
5–Plateau of In Rakli (M'Zarag)
6–our tent camp in the zone of soil pulverization
7–Plateau of the Assekrem (2,727 m.–8,948 ft.)

well-preserved relictic mull horizons of earthy Braunlehm

earthy Braunlehm

nonpulverized relicts of earthy Braunlehm, corresponding approximately to the border line of soil pulverization

severely eroded steep slopes without soil covers, or bare rock walls

(Landscape contours after P. Bordet)

164

In some profiles in situ even the humus horizons are preserved, and these could not have been produced under the present environmental conditions because of the lack of a closed plant cover. The best humus horizons are found in a closed area at the pass level between the Assekrem and In Rakli mountain (Figure 14), below the hermitage hut of the famous missionary Père de Foucauld on the slope of the Assekrem at an altitude of 2,650 m. (8,690 ft.). The A horizons have a thickness of about 20 cm. and are dark gray-brown in a wet state; the humus form is mull. The (B) horizons have a thickness of 20 to 76 cm., are ocher-brown in a wet state and are considerably denser than the A horizons.

In approaching the mountain valleys on the lower parts of the slopes, C_m horizons were observed, instead of the (B)/C horizons. They were gray, with precipitations of manganese hydroxides in the shrinkage cracks. The basalt structure of these horizons was well preserved, although they had been transformed into a completely soft and easily cuttable mass. The texture in all horizons was a clay loam.

MICROMORPHOLOGY OF THE ATAKOR BRAUNLEHM RELICTS

Thin sections of the Braunlehm relicts of the Atakor of the Ahaggar reveal the fabric of an earthy Braunlehm which, according to our comparative investigations, was formed in the montane belt of the tropics with its fog woods. A strong earthening was observed in the A horizon. The flocculated amorphous iron hydroxides were deep brown. The soil mass was loose and had a fine sponge fabric and partly rounded aggregate formation. The relict humus displayed good decomposition and humification; the humus substances had been worked completely into the clay substances. The (B) horizons had a more dense matrix with a brown granulation. The iron hydroxide granules were partly flocculated. Under crossed Nicols, double refractive fabric parts, produced by particle arrangement, were observed in some cases. Some dark-brown, well-rounded concretions of iron hydroxide accentuate the Braunlehm character. They are more common in Braunlehm sediments than in Braunlehm profiles in situ.

Thin sections of the parent material revealed a basalt whose groundmass of plagioclase ledges contained numerous small magnetite grains, larger phenocrystals of olivine and some titanium-containing augites. In the C_m horizons the plagioclases and augites were softened and had lost much of their birefringence. The olivines were transformed into reddish bowlingite. In the soil horizons all minerals were completely weathered with the exception of the magnetites, which retained, unchanged, their crystallographic delimitation.

According to R. C. Mackenzie and W. A. Mitchell (Aberdeen), who kindly took charge of investigating the clay minerals and the iron hydroxides, the matrix of the (B) horizons uniformly contained about 25 per cent kaolin of the fireclay type. From the differential thermal analysis a further component of illitic nature was recognizable. The x-ray analysis gave 40 per cent illite-vermiculite, 5 per cent montmorillonite and 5 per cent quartz. Iron hydroxide mineral content could not be determined by either of the two methods, but this was expected since iron hydroxides in Braunlehm are amorphous. The relatively high content of illitic clay minerals by comparison with fireclay minerals is explained by the altitude and by the fact that the latitude in which these relicts are situated is almost at the border of the subtropics. The vermiculite component is probably due to the presence of allochthonous materials in which biotite was often a constituent.

BORDER LINE OF SOIL PULVERIZATION

The pulverization of relicts of humid tropical soils is caused by the desert climate. The border line, where pulverization of soils ceases, is therefore of particular interest for soil investigations of desert mountains. Pulverization was so strong at our camp on the edge of the great volcanic circus of the Assekrem at an altitude of 2,000 m. (6,560 ft.) that we could not enter our tents without kicking up suffocating clouds of dust. Two thousand feet further up, at the altitude of the pass (the habitat of the relicts of A horizons), pulverization was hardly noticeable (Figure 14). This accounts for the preservation of the humus horizons at these altitudes. There was no pulverization in the vicinity of the plateau of the summit of the Assekrem. The border line for pulverization was therefore an altitude of about 2,200 to 2,300 m. (7,215 to 7,545 ft.).

In the sediments of the pediment and hammada formations Braunlehm character greatly predominated. Braunlehm sediments formed the surface layer, whereas deposits of a different nature covered the lower slopes or filled depressions. In level positions the sediment surfaces always displayed erg formations, with slightly rounded stones covered by a blackish desert varnish.

The material of the sediments was by no means homogeneous; although most of the deposits had the E/(B) character of Braunlehm, there were so many clods of other kinds that a sort of mosaic was formed. Sometimes E/(B) clods of Rotlehm were found, and even small clods of a former A horizon of a Braunlehm. The Braunlehm character was much better preserved in these soil sediments than in the Braunlehm profiles in situ; the latter displayed a more advanced earthening. This soil turned into very sticky mud masses during the frequent rains on our excursions. All these peculiarities make the differences between the Braunlehm sediments and the surface layers of the Braunlehm soils in situ very clear.

THE FOSSIL ROTLEHMS OF THE ATAKOR

The Rotlehm varieties that we observed in the Atakor of the Ahaggar occurred only as fossil soils. They exhibited different characteristics according to their age. The youngest were bright red, yet, microscopically, had more or less the character of a slightly rubified Braunlehm. In contrast, there were older varieties that displayed advanced red earthening, wherein the precipitated iron hydroxides formed a water-stable sponge fabric rich in pore spaces. These earthy Rotlehms displayed a considerable loss of silica as a result of heavy leaching (analyses cited by P. Bordet, 1952). Their fabric plasma, according to the investigations of R. C. Mackenzie and W. A. Mitchell, was considerably richer in kaolinite (50 per cent) than the Braunlehm of the Atakor. Illite was present in the amount of 5 to 10 per cent. The content of crystallized iron hydroxide was 18 per cent. After a half hour of heating to 600°C. the x-ray analysis showed about 30 per cent hematite.

COMPARISON OF THE ATAKOR BRAUNLEHM WITH RECENT BRAUNLEHM

A recent equatorial Braunlehm of Fernando Póo (3° N) formed on the same parent rock was selected for comparative investigations (Figure 15). The matrix of the (B) horizon contained a large amount of fireclay mineral but almost no illitic component. In the montane belt of the fog woods the equatorial Braunlehm exhibited a similar earthening. From the similarity of the profile formations and micromorphology it can be concluded that the Braunlehm of the Atakor of the Ahaggar was formed under the influence of an ever-humid, subtropical climate (also subject to the influence of the altitude) under a dense forest cover which had reached the highest elevation of the mountains (Figures 15, 16 and 17). Julius Büdel, who traveled in the Atakor a little later, estimated, on the basis of my

Figure 15. Profile of the Earthy Braunlehm Relict of the Atakor of the Ahaggar Mountains of the Sahara Compared with Soil Profiles in Other Regions of Africa

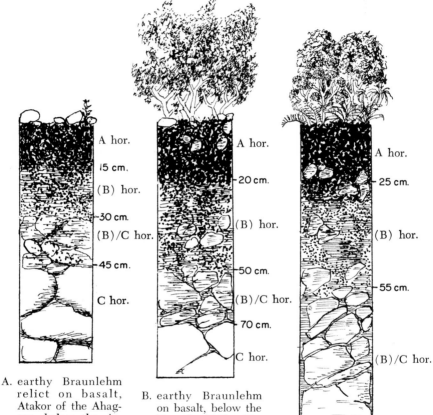

A hor.

15 cm.

(B) hor.

-30 cm.

(B)/C hor.

-45 cm.

C hor.

A. earthy Braunlehm relict on basalt, Atakor of the Ahaggar, below the Assekrem, at 2,650 m. (8,690 ft.)

A hor.

-20 cm.

(B) hor.

-50 cm.

(B)/C hor.

-70 cm.

C hor.

B. earthy Braunlehm on basalt, below the Pico de Teide, Tenerife, Canary Islands, at 1,830 m. (6,000 ft.)

A hor.

-25 cm.

(B) hor.

-55 cm.

(B)/C hor.

-120 cm.

C hor.

C. earthy Braunlehm on basalt, Fernando Póo, Gulf of Guinea, at 1,980 m. (6,500 ft.)

Figure 15 (cont'd)

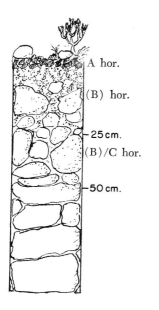

A hor.

(B) hor.

25 cm.

(B)/C hor.

50 cm.

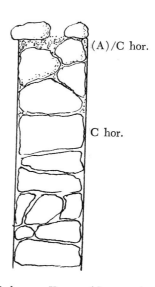

(A)/C hor.

C hor.

D. subtropical Dry-Braunerde, Tenerife, Canary Islands, at 90 m. (300 ft.)

E. brown Yerma (desert raw soil) on basic volcanic ash, La Sal, Cape Verde Islands, at 40 m. (130 ft.)

mull fabric (humid variety)

earthy braunlehm fabric

mull fabric (dry variety)

dry-braunerde fabric

comparative investigations as well as investigations of his own, that the yearly precipitation during the time of the Braunlehm formation was 1,000 mm. (40 inches). The approximate homogeneity of the precipitation throughout the year is indicated by the character of the soil.

THE PRESERVATION OF THE ATAKOR BRAUNLEHM RELICTS

A strong transformation of Braunlehm on basalt is produced by an alternating humid and dry climate with effective dry seasons; the result is an intense rubification followed by red earthening. Such climatic conditions did not exist after the development of the Braunlehm cover in the Atakor. However, the climate was an alternating humid and dry climate at the time the Rotlehm varieties of the hill belt were formed. This belt had been covered by lava flows; but the soils, judging from their characteristics, reflect an earlier period of soil formation in which environmental conditions were quite different from those that exist at present in the area.

The early loosening and pulverization processes were effective only on the surface layers of the Braunlehm relicts, and then only at lower and middle altitudes. Beginning at about 2,200 to 2,300 m. (7,215 to 7,545 ft.), the intensity of pulverization and wind erosion decreased and these processes ceased completely at greater altitudes.

The brown earthening transformation produced by annual differences in temperature and humidity increased under the present environmental conditions, particularly during the frost periods of winter. However, this kind of transformation had the effect of preserving the soil rather than destroying it.

None of the later climatic changes was effective enough to change the character of the earthy Braunlehm, not even the present desert climate in conjunction with high altitude. In spite of the state of preservation of the Braunlehm cover, the landscape is one of destruction, and the residues of vegetation and fauna of the earlier humid periods are scanty.

AGE OF THE ATAKOR BRAUNLEHM RELICTS

Since nowhere in the Atakor of the Ahaggar have the Braunlehm relicts been covered by extensive basaltic lava flows, they antedate the volcanism. Furthermore, the Braunlehms are separated from the older fossil Rotlehms by these lava flows or do not intergrade into the Rotlehms where they had been laid bare by erosion. I agree with Büdel (1955) that the humid period of the Braunlehm in the Atakor must have been contemporary with an ice age in the present-day temperate zone. Soils

Figure 16. Recent Soil Belts on Basalt of Some Mountains of Africa and the East Atlantic Sector

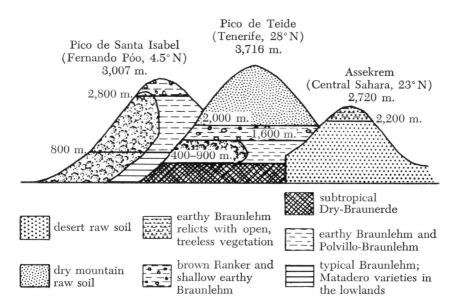

Figure 17. Soil Belts Formed on Basalt during the "Last Great Pluvial," Reconstructed from the Soil Relicts

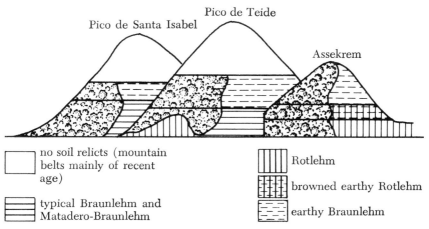

171

represent the most unequivocal evidence of climates; and in our warm age in the temperate zone, which is clearly a repetition of the warm inter-glacial ages of the Pleistocene period, a definite increase in the xeromor-phism of the soils is discernible almost everywhere in the subtropics and semihumid tropics. This xeromorphism is advancing towards the polar regions, as well as towards the Equator, and to belts of higher altitude in the mountains. I myself have observed this phenomenon in the East Atlantic islands, particularly in the lowland zone and hill belt of the Canary Islands and Cape Verde Islands.

As I have already pointed out, the presence of Braunlehm relicts that were never covered by lava flows in the Atakor indicates that the volcan-ism cannot be of recent age. The fairly undisturbed geomorphology of the region suggests that it must have corresponded very closely to the youngest ice age, the Würm of the Pleistocene epoch. Since it is possible to distinguish at least two or three different fossil Rotlehm covers in the lower belts, separated by lava flows of different ages, these might corre-spond to the older glacial periods in the temperate zone of today. These Rotlehm covers are different from the Rotlehm relicts on crystalline rocks of the plain of Tamanrasset.

AGE OF THE ROTLEHM RELICTS OF TAMANRASSET

In the plain of Tamanrasset south of the mountains of the Atakor, resi-dues of a Rotlehm occur everywhere on the outcropping gneisses, amphib-olites and mica-schists. We found profiles in situ of the (B)/C and (B)/C$_w$ (white horizon) varieties or Rotlehm sediments. Their matrix is primarily kaolinitic; the (B) horizons have for the most part been removed by erosion, leaving the very dense (B)/C and C$_w$ horizons behind.

Since the Rotlehm relicts on crystalline rocks are partly covered by lava flows, they antedate the volcanism of the area. They correspond to the Rotlehm relicts not only in other parts of the Sahara but also in other parts of nontropical Africa. Corresponding varieties can be found in differ-ent parts of the Mediterranean zone. These Rotlehm relicts represent soil formations whose present border line is situated considerably more to the south; this line, however, was moved northward on a broad front during an earlier humid period of long duration. This humid period, which lasted longer than the humid periods of the Pleistocene by many millions of years, is placed at the earliest in the Tertiary, when the border of the tropics ran much farther north, without fluctuation.

XXII

Soils of the Mediterranean Zone

The European countries around the Mediterranean Sea, although they form a transition zone between the subtropics and the temperate regions, represent a soil province of distinctive character by reason of the soil development, the plant cover and fauna and the types of agriculture. The European Mediterranean is separated from neighboring regions by the high mountains of the Pyrenees, the Alps, the Balkan Mountains and the Atlas; and, in the south and east, by the deserts or desert steppes of the Sahara, of Arabia and of Iran. It comprises the three European peninsulas, as well as Asia Minor and the coastal regions of North Africa. In particular it consists of Spain, Portugal, southern France, Italy, Morocco, northern Algeria, Istria, the southern Balkan countries, Greece, the Crimea, the coastal areas of Asia Minor and the western slopes of the Caucasus.

In other parts of the Northern and Southern Hemispheres, moreover, we recognize regions that are very similar in that their environmental conditions correspond closely to those of the Mediterranean. Such regions are central and southern California, central Chile, the South African Cape of Good Hope Province, Iraq, Iran, Turkmenia and southern and southwestern Australia.

The Mediterranean region and regions like it have a great variety of landscapes, from semideserts, which have the lowest annual precipitation of Europe, to high mountain forests with a precipitation of 3,000 mm. (120 inches) or more. Each region has distinctive soils, plants, animals and human population.

ZONATION OF THE MEDITERRANEAN

With respect to environmental conditions and soil development, the Mediterranean countries divide clearly into a cooler and more humid northern zone and a warmer and dryer southern zone. In Europe this zonation is very clear in the Iberian Peninsula as well as in the Apennine and the Balkan peninsulas. It is also very evident in Asia Minor, in Iraq and in Iran. In regions with considerable continentality there is also a middle zone in which the approach to subtropical conditions is much less marked than in the southern zone. The northern zone contains soil formations that are transitional to Central European soils, although there may exist essential differences, as in the case of Atlantic Ranker, Podzol Ranker and Asturian Podzol. The southern zone, like the belt of the hardwood forests, possesses soils of a distinctive character, although a series of formations resembles African varieties. In this zone more salt soils can be observed. Most of the southern border of this zone intergrades into semideserts and deserts like those in North Africa and regions south of the Atlas, into the Great Sahara.

HYPSOMETRIC ZONATION OF SOIL DEVELOPMENT IN THE MEDITERRANEAN

With increasing altitude in both main zones of the Mediterranean, very typical differences exist in regard to profile development and micromorphology, although there are never as many belts as in the temperate zone. On silicate rocks the mediterranean Dry-Braunerde of the middle and southern zone intergrades at an altitude of about 1,000 m. (3,300 ft.) into semihumid to humid Braunerde varieties; these are followed by a well-developed belt of subalpine and alpine Rankers (depending on the nature of the parent rock) and end with a belt of alpine Råmarks. In the northern zone humid Braunerde is present in the hill belt, intergrading into Semipodzols in the mountain belt. Podzols are comparatively rare and usually occur on specific parent rocks; for example, Asturian Podzol is a formation on siliceous sandstone in northern Spain.

RECENT SOILS ON SILICEOUS AND SILICATE ROCKS

In the Mediterranean the difference between recent and relict soils is easy to recognize, since relict soils are much better preserved than in the temperate zones. This is particularly true of the Rotlehm relicts, soils that are preserved to such a degree that many investigators in the past believed that the soils most typical of a mediterranean climate were red. Up to now it has been very difficult to be sure that a red soil was of recent age, whereas it has been easy to establish that many Rotlehm and even Terra rossa varieties are of Tertiary or Pleistocene (interglacial) origin. The present soils, formed by the much dryer environmental conditions of the present, are characterized by low chemical weathering and low accumulation of clay and free iron hydroxide and are primarily light brown. Xeromorphous transformation of former humid soils mainly affects their physical properties and only to a small degree their color. The most important soils on siliceous and silicate rocks are discussed below.

MEDITERRANEAN DRY-BRAUNERDE[*]

Whereas humid Braunerde with excellent mull humus under a closed plant cover is easily formed on silicate rocks in the humid temperate zones, the soil formed in the middle and southern zone of the Mediterranean region on the same rock under the present environmental conditions is similar to a Braunerde in profile and micromorphology but different from it in that it is light in color, is subject to little chemical weathering and has a low humus content, thin humus horizons and much less soil life. The vegetation on uncultivated mediterranean Braunerde is a sparse and open xerophilous plant cover dominated by dwarf shrubs of the holm oak, *Quercus ilex*. This soil was originally a forest soil, mainly under stands of *Quercus ilex* trees, but today it is found only in the form of a few small forest relicts and is dried out most of the year.

The progressive xeromorphism of the mediterranean Dry-Braunerde is in large degree a consequence of human activity. Investigations in central Spain have shown that the soil transformation began during the Roman colonization as a result of the Roman type of farming. Today the soils respond to farming with a rapid and catastrophic decline of fertility. The causes are desiccation of the soil mass in consequence of the strong radiation of the Mediterranean summer sun and the lack of shade-giving vegetation, decrease in the binding power of the colloids and clay minerals, loosening or complete breakdown of the soil fabric, increasing dust formation and strong erosion of the superficial soil layers by wind and flowing water, underdevelopment of the vegetation, decrease in soil life and humus formation, decrease in chemical weathering and therefore decreases in clay formation and mobilization of plant nutrients (particularly available potassium), low moisture retention and insufficient water supply for plant growth.

I had occasion to investigate the transformation of the mediterranean Dry-Braunerde in its different stages by analyzing soils associated with

---

[*] In view of the fact that the concept of this soil applies wherever it occurs in other parts of the globe, including the Southern Hemisphere, the term "meridional" Braunerde had been replaced by mediterranean Dry-Braunerde. Another name for this soil is brauner Trockenwaldboden.

forest relicts, in fruit groves and on grain farms. The most important fruit groves on silicate rock in central and southern Spain are those of *Quercus ballota,* which bears edible acorns that are sold in the markets of southern Spain. Under the shade of these oak groves, there developed closed grass pastures which for centuries offered excellent pasture and acorn mast for pigs. These groves had an excellent effect on the soil. In those areas where the custom of pasturing pigs in these groves has persisted to the present, the soil layer is remarkably thick and displays a good earth formation, good biological activity, favorable humus and aggregate formation and a well-balanced water economy.

Just the reverse is characteristic of soils which have been consistently planted to crops. Unfortunately, the great need for arable land in the mediterranean area induces landowners to cut down the groves and to till the soils. With the rapid loss of fertility and visible impoverishment, these soils become barren and turn into wasteland in only five to ten years.

Mediterranean Dry-Braunerde covers large areas within the middle and southern subzone of all mediterranean regions of the globe underlain by silicate rocks. It presents great problems for farms that do not have irrigation systems. Irrigated sediments of that soil, or mediterranean Braunerde-Vega, are fertile, but this soil is not widely distributed.

Mediterranean Dry-Braunerde has a fabric intermediate between that of a Braunerde and a Braunlehm, but it is not well developed and the weathering of the mineral grains of the sand fraction is weak. There is some aggregate formation and flocculation of free iron hydroxide, but the binding power of the soil colloids is low and the aggregates have a tendency to break down into smaller fragments or even to pulverize.

HUMID AND SEMIHUMID BRAUNERDE VARIETIES

Humid Braunerde on silicate rocks is the predominant soil formation of the northern mediterranean subzone. It is very similar to the Central European Braunerde. It is easily worked, very water stable, well flocculated and has a well-developed sponge fabric. Stebutt referred to this Braunerde variety when he stated that the best-developed and most stable Braunerde occurs in the extreme southern part of the (humid) Braunerde zone (Stebutt, 1930).

Semihumid Braunerde occurs in many places in the mountain belt of the middle and southern mediterranean subzone. It represents the climax formation of this belt on silicate rocks.

SILICATE RAW SOILS AND XERO-RANKERS

On the border of the southern mediterranean subzone next to the desert or in islands that are very dry, humus formation, chemical weathering, soil life and colonization by higher plants decline further in intensity, with the result that xeromorphism is considerably aggravated. The soils do not form a (B) horizon (Xero-Ranker), or they have neither a (B) horizon nor an A horizon (raw soil of the dry desert, Yerma). Xero-Ranker is low in humus content, has little soil life and supports few higher plants.

MEDITERRANEAN MOUNTAIN RANKER

At altitudes above 1,000 m. (3,300 ft.), well-developed Ranker varieties in the form of montane, subalpine and alpine soils are widespread. They are fairly rich in humus, and therefore are dark brownish-gray to blackish-gray. They occur in the form of pre-stages in the realm of semihumid and humid Braunerde or alpine Sod Braunerde. They can be distributed over extensive mountain areas and may even occur as climax formations. Such Ranker varieties may be found in connection with tangel humus; and as Tangel Ranker, they are very characteristic of some Spanish high mountains on granite and gneiss (Kubiëna, 1953a). Or they may occur in connection with mull-like moder or mull humus as alpine Ranker (Humus-silikatboden). Mull-Ranker in the granite mountains of Spain and Portugal may develop mull horizons up to 102 cm. (40 inches) thick; this form occurs on the coast of Galicia even in the hill belt (Atlantic Ranker) (Franz, 1956).

SEMIPODZOLS AND PODZOLS

In the southern mediterranean subzone, Semipodzols and Podzols become rare but nevertheless are sometimes present at altitudes which correspond to them in the temperate zones. Semipodzols and Podzols are generally replaced by humid Braunerde, Tangel Ranker and alpine Ranker. Well-developed Podzols, which are among the principal soil formations at altitudes of 1,800 to 2,500 m. (5,900 to 8,200 ft.) in the Alps, are rare in the northern mediterranean subzone (in northern Spain they occur only on siliceous rocks and not on silicate rocks).

ASTURIAN PODZOL

The Asturian Podzol of northwestern Spain occurs in the northern subzone in a rather warm climate for Podzol formation with raw humus. The province of Asturias has a humid oceanic climate, and there are many chestnut trees and racemose oak trees. These Podzol soils, which extend almost to the coast, occur with specific parent rocks and vegetation; they develop primarily on Devonian siliceous sandstone or stony moraines (with sandstone débris) and a plant cover similar to that found with the Podzols of the Cape of Good Hope. This vegetation consists primarily of *Erica* varieties, among which the Atlantic species *Erica ciliaris* is the most common in northern Spain.

The Asturian Podzol soils are distributed mainly on the deforested, heath-clad hilly spurs of the Cantabrian Cordillera at altitudes of 300 to 1,000 m. (1,000 to 3,300 ft.). They develop from a dystrophic Ranker or, under the same or a similar plant cover, a so-called Podzol Ranker (Kubiëna, 1953a). This soil develops a bleached sand layer of a whitish violet-gray to bright-gray color under an A_0 horizon of raw humus or coarse moder. An A/B horizon is missing, but leached humus substances and free iron hydroxides are precipitated in blackish to reddish-brown stains in the cracks and fissures of the parent rocks. In the Cantabrian Mountains, Podzol Ranker occurs only in small habitats, but in the foothill belt and in the vicinity of the coast, particularly in eastern Asturias, it may occupy considerable areas.

The Asturian Podzol, as it is outlined in the pre-stage of the Podzol Ranker, develops in the form of an Iron Humus Podzol. Its most striking characteristic, however, is that it develops a white, chalk-like accumulation layer of colloidal silica (Si horizon). The matrix of this horizon is friable, for the silica is precipitated in loose, finely dispersed masses between the quartz grains of the fabric skeleton. On the other hand, the silica in the fresh sandstone acts as a very effective binding substance. The rock is hard and tough, and breaks conchoidally into splinter-like fragments with sharp edges like flint. The weathering peculiar to this sandstone is partly visible in the C_1 horizon or in rock fragments in the soil and on the soil surface. The hard silica is softened, the stone mass becomes fragile and can be broken easily by the hands. What causes the silica to dissolve and leach through the acid A_e horizon to accumulate below the B_s horizon cannot be explained. The influence of the warm, humid environmental conditions is evident.

The A_0 horizon of the Asturian Podzol has a marked tendency to decompose, primarily through the action of small arthropods. The humus layers are well developed and deep black. Well-decomposed raw humus is transformed into a humid moder consisting of abundant accumulations of small animal droppings. The bleached sand layers are 10 to 15 cm. (4 to 6 inches) thick and grade into a blackish-brown orterde layer (B_h horizon). The humus sols that produce the blackish colors derive from the animal droppings of the A_0 horizon. This can be clearly recognized in thin section analysis, since one sees all transitions from peptized sols to recognizable residues of partly preserved fragments of animal excreta in the intergranular spaces of the B_h horizon. In the B_s horizon there is no coating on the quartz grains, but rather flaky deposits, rich in iron hydroxide, are visible in the intergranular spaces. In the blackish humus veins the fabric of the rock is well preserved. Here the silica is removed and replaced by iron-containing humus substances.

SOUTH AFRICAN CAPE PODZOLS

These are also Iron Humus Podzols under dense *Erica* heaths in a climate remarkably warm for a heath Podzol region. They cover the rock formations of the Cape of Good Hope. As in Asturias, the parent material is a siliceous sandstone. However, in spite of the similar conditions and many analogies in profile development and micromorphology, the Cape Podzols do not develop an accumulation layer of silica. This is explained by the fact that the siliceous binding substance in its sandstone has to a much greater extent been transformed by aging into quartz. Even some Humus Podzols are formed, particularly where there are trees.

The Iron Humus Podzol supports a great variety of *Erica* species. Dystrophism is widespread. Even in the botanical garden of Kirstenbosch, the water in the creeks is brown. Nevertheless, profiles of bleached Braunlehm can be found on Cape sandstones that contain a good deal of clay (the slopes of Table Mountain). Because of their great erodibility, these Braunlehm soils, bleached or unbleached, easily form thick colluvia of Braunlehm sediments. On the plateau of Table Mountain, humid, mull-like Ranker to Proto-Ranker occurs on the little-decomposed rock surfaces. On clay schists above the Funchhoeck Valley from about 300 m. (900 ft.), humid Braunerde similar to that of Asturias is formed on the same parent rock.

THE RELICT AND FOSSIL SOILS

Paleosoils play an important rôle in all mediterranean regions. Relict soils especially may cover considerable areas. Varying from intense ocher to dazzling red or orange, the paleosoils are much more striking in color than the recent soils. Profiles in mountain areas are rarely complete and occur mainly as residues of erosion. Frequently only the (B)/C horizons are preserved above the parent rocks. Recognition of a paleosoil is also facilitated by the fact that such a soil cannot be formed again in places where it has been eroded; it can only be replaced by recent soils, such as mediterranean Dry-Braunerde in the foothill belt, semihumid Braunerde in the mountain belt and alpine Sod Braunerde in the alpine belt of the middle and southern subzones. Another peculiarity of the paleosoils is that they usually occur in thick deposits of soil sediments, whereas deposits of recent soils of any considerable thickness are almost non-existent. Besides relict profiles in situ in the mountain area, the corresponding soil sediments can very often be found in the valleys. By thin section analysis it can be determined which materials belong together and which must be considered of different origin.

Comparative research with soil materials of the tropics and subtropics affords the possibility of finding out what the environmental conditions were in earlier periods. The state of preservation of the residues of the paleosoils is, in general, much better in mediterranean areas than in the temperate zones. Particularly in the middle and southern subzones, the factors that cause transformation are not effective enough to change the character of the older soils fundamentally.

Rotlehm relicts are by far the most extensive of the relict soils of the mediterranean regions. There are numerous small erosion remnants no more than 1½ to 3 m. in diameter, but there are also closed areas of relict soils in situ with a diameter of several kilometers, even on granites, gneïsses and Paleozoic clay schists. The largest areas of Rotlehm are found in the forest region of La Selva near Gerona in Catalonia, in a granite section which has never been cultivated. Another Rotlehm area known as La Vera is situated on the south slopes and the adjacent territory of the Sierra de Gredos in central Spain. In the vicinity of Thessaloniki in northern Greece, there is an area of Rotlehm on chlorite schists, and in Extremadura, western Spain, there is one on Paleozoic clay schists. The Rotlehm of Thessaloniki is sold commercially as potting soil.

The thick Braunlehm and Rotlehm sediments are very important in all mediterranean countries, just as are the soil relicts in situ. Some cover large parts of a district; they generally predominate in fluvial deposits. Paleosoils in situ as well as soil sediments are also common in the form of fossil formations. They are good indicators of age and former environmental conditions.

AGRICULTURAL IMPORTANCE OF THE RELICT SOILS

The relict soils display a degree of weathering, of clay accumulation and of maturity that cannot be attained under present environmental conditions, with the extremely dry summers. These soils are therefore far more valuable to man, more fertile and more balanced in their water relations, than the recent soil formations. They are very important for agriculture. It is still possible to preserve some of them, but if they are destroyed, they can never be replaced. The protection of relict soil areas is therefore an important land conservation issue in the dry areas of the mediterranean countries.

HUMAN INFLUENCE ON EROSION OF RELICT SOILS

Geologically, the relict soil areas are of special interest, for they sometimes represent entire sections of landscapes of the geological past that have remained essentially intact. Even where only isolated small profile remnants are preserved, it may be possible to reconstruct the elevation and extension of the former land surface.

From the position of the relict soils and their relation to areas of early colonization, from the density of settlement and observation of the progress of erosion and its causes at the present time, it is clear that the destruction of the relict soil covers has been very seriously accelerated by man in the Mediterranean area.

In Spain the worst erosion of the relict soils has taken place precisely where Roman colonization was densest. It is a striking fact that the Rotlehm relicts in the granite area of Los Pedroches in the Sierra Morena of southern Spain are preserved on the steep middle slopes, whereas they are completely eroded on the more level lower slopes. The steep slopes were never used as arable land and therefore were protected by a dense plant cover of forest vegetation or bushes of *Cistus* or *Erica* varieties.

PALEOPEDOLOGICAL SIGNIFICANCE OF THE FANGLOMERATES

For the paleopedologist, knowledge of the erosion sediments of the paleosoils is as important as knowledge of those soils in situ. Not until a complete picture of the entire cycle of soil development, soil erosion and deposition of the soil sediments is obtained is it possible to reconstruct the paleopedology of a period satisfactorily.

The Paleozoic clay schists are materials which have a strong tendency

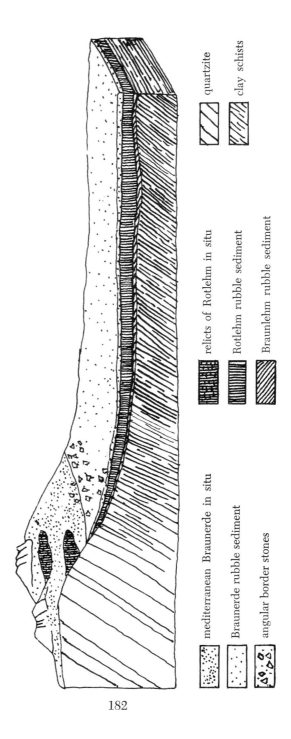

Figure 18. Cross Section of a Raña

quartzite

clay schists

relicts of Rotlehm in situ

Rotlehm rubble sediment

Braunlehm rubble sediment

mediterranean Braunerde in situ

Braunerde rubble sediment

angular border stones

182

to form fanglomerates, that is, deposits of stone mudflows that originate in dry climates under the influence of heavy rains on a soil surface little protected by vegetation. There are extensive deposits of fanglomerates in the Paleozoic schist areas of central Spain, where they are known as *rañas*.

Rañas generally set in at the elevated quartz ridges which stick out from the *rumpf* plane of the clay schists. They begin in a thin layer, and gradually cover the valley plain in meseta-like shapes to a thickness of 45 m. (150 ft.) or, in extreme cases, 1,000 m. (3,300 ft.). The stones of the raña next to the ridge are angular to moderately rounded and alternate with large blocks. With increasing distance from the quartz ridge, the degree of breaking up and rounding increases. The material is not stratified, yet it is not homogeneous either, for it displays distinct differences with increasing depth. One may therefore conclude that the rañas were formed not in a single period but in several different periods, in which different soil materials were available for the formation of the rañas. Since these materials correspond to the soil covers that prevailed in each period, it is possible to determine the distribution and the succession of the different types of soil formation (Figure 18).

MICROMORPHOLOGICAL INVESTIGATION OF A RAÑA

A micromorphological analysis of the materials at different depths of one raña was made (Kubiëna, 1958). The stony and sandy surface layer, light brownish-gray in color (Munsell values from light brown 10 YR 6/3 to light brown 7.5 YR 6/4) and about 50 cm. (20 inches) thick, contained material of raw soils (Syrosems) and mediterranean Dry-Braunerde.

Underneath was a stony, clayey layer about 4 m. (13 ft.) thick and red in color (Munsell red 2.5 YR 5/8) which contained only material of earthy Rotlehm with flaky precipitations of iron hydroxide.

The lowest unconsolidated layer consisted of a dense clay with braunlehm fabric that was ocher in color (brown-yellow 10 YR 6/8). This Braunlehm material showed some gleyzation (Munsell gray 2.5 YR 6/8) and pseudogleyzation, and together with the clay schists it formed the impervious basis of the raña.

A similar composition has been observed with all rañas investigated. The highest layer of the material of mediterranean Dry-Braunerde continues from the lower positions on the hill slopes directly to higher land where recent soil formation is taking place.

FORMATION AND AGE OF THE RAÑAS

All recent investigators agree that the formation of the Spanish rañas took place during the late Tertiary. Originally closed, the mud and rubble covers were later divided into different tables by the deep-cutting post-Tertiary erosion (Hernández Pacheco, 1949). In consequence, the valleys of the large rivers intersected all the layers underneath, down to the Paleozoic substratum. Thus, the rañas are composed of the erosion deposits of the Tertiary soil covers, inclusive of their (B)C horizons (hence the deposition of the great masses of weathered stones and rubble).

From the sequence of the different layers of the rañas it can be concluded that the first erosion took place in a period in which the soil cover in situ was composed almost entirely of Braunlehm varieties (Kubiëna, 1955a). This was followed by a period with much heavier erosion and the deposition of mighty mudflows consisting of soil material that was chiefly Rotlehm in character. The formation of the rañas terminated with the deposition of erosion sediment that had the character of mediterranean Dry-Braunerde or of raw soil, the deposition being produced by environmental conditions very much like those of the present time.

The sequence of soil sediment layers corresponds to the climatic change in the late Tertiary. Whereas an ever-humid, warm climate must be assumed for the early Tertiary in this region, a humid tropical climate must be assumed for the Eocene; a change to an alternating humid and dry savanna climate took place in the Oligocene and attained its highest development in the Miocene. In the Miocene, however, great areas of Braunlehm covers must have been transformed by rubification very gradually. The predominant soil formation developed more and more as a highly rubified Rotlehm. But deeply weathered, iron-rich Rotlehm varieties with a high clay content cannot be formed in a desert or semidesert climate (Kubiëna, 1955a). In the Miocene, the alternating humid and dry climate had already been repeatedly interrupted by dry periods in which sporadic rains in the form of heavy showers produced heavy erosion and the formation of mudflows and rubble flows as they drenched the areas bare of protecting vegetation. But Rotlehm formation resumed between these interruptions, and it must be assumed that it continued in the early Pliocene. Not until the late Pliocene was there a shift to a cooler, dry climate. The removal of the thick Rotlehm cover continued, and, with respect to new soil formation, silicate raw soil and mediterranean Braunerde developed, instead of humid soils; characteristically, the new soil formations were subject to little weathering and had a low content of clay and free iron hydroxide.

AGE OF THE SOIL RELICTS IN SITU

In addition to the soil sediments in the rañas there are some profile remnants of the paleosoils in situ in protected locations. Their micromorphology corresponds to that of the sediments in the rañas. Since all the formations were produced on the same parent material (Paleozoic schists), it is evident that the environmental conditions were very different. Inasmuch as there are no fossils in the rañas, the dating is difficult (Oehme, 1935, 1942; Hernández Pacheco, 1949). Paleopedologically, the culminating point of Rotlehm formation in situ must have been in the Miocene, and the Braunlehm formation (relicts are much rarer) in the hill belt of this area must have taken place prior to that time, or in the early Tertiary. The formation of the rañas could not have been completed in the Miocene because the sediments of the highest layers, consisting of material of silicate raw soil and mediterranean Dry-Braunerde, make this dating difficult. It is very probable that Rotlehm formation continued far into the Pliocene, and that the dry soils began to predominate only towards the end of it. Some remnants of Rotlehm profiles in situ can still be found on the Pliocene basalts of Ciudad Real in south-central Spain. In the nearby valley of the Río Bullaque, in places where the raña had been covered by basalt flows, the raña did not have a light-brownish-gray surface layer of Dry-Braunerde sediments but terminated instead with sediments of Rotlehm character.

HUMID RENDZINAS OF THE MEDITERRANEAN

These Rendzinas are young grassland or forest soils of the northern mediterranean subzone or grassland soil of the high mountains in the middle or southern subzone. Since they are subject to heavy leaching of calcium carbonate, only clastic calcite remains in the fabric of the dark humus horizons, but there are well-developed calcium horizons under the A horizons.

Sometimes humid Rendzinas also occur in coastal areas in the middle and southern subzones, where the precipitation is greater locally. For example, on the Peñon Ifach, a limestone cliff above the Mediterranean near Alicante, humid Rendzinas have developed on the side exposed to the sea, whereas very dry Xero-Rendzinas occur leeward.

RAW SOILS AND XERO-RENDZINAS

In the lowland area and hill belt of the middle and southern mediterranean subzones, the process of soil development is quite different from that in the northern subzone and in the high mountain areas. Whereas a raw soil phase is almost never found in the northern subzone because of the intense and rapid humus formation, this phase is well developed and has a characteristic development phase in more southerly zones, particularly in the middle zone. Because of this, carbonate raw soil formation (carbonate Syrosem) is common, primarily on marl loams and marl rocks. These are parent materials on which a well-developed Mull-Rendzina is easily produced in humid areas. Since a humus horizon is lacking in the dry regions, such Syrosems have often been called white Rendzinas. They have little effectiveness biologically and act almost to preserve instead of decompose and humify organic residues. Since they occur in the dry habitats of the middle and southern subzones, they are poor in vegetation, but still they are suitable for olive groves. In depressions, on gentle slopes and in old erosion gullies, some light-gray humus formation is seen.

The Xero-Rendzinas of these dry areas are low in humus and are therefore always a very light gray. Their surface layer contains large amounts of recrystallized calcium carbonate, and this makes the color of the humus layer even lighter. In dense marls the accumulation of this recrystallized calcite may, under special conditions, cause subsoil cementation (chalk caliche) or superficial lime crusts (croûtes calcaires) or a rock-like hardening of almost the whole soil layer. In gypseous marls gypsum crusts may be formed.

Xero-Rendzinas on marls generally exhibit mull humus, but because of the sparse vegetation and the poverty of organic raw materials the humus accumulation is small. Nevertheless, intense activity of the earthworm fauna can be observed during the short rainy periods. Earthworm activity ceases completely only in raw soils, on hard parent rock with thin soil covers.

HUMID TERRA FUSCA AND TERRA ROSSA

Humid Terra fusca, a generally humus-deficient, decalcified, clayish soil with dense braunlehm fabric, is common in the limestone areas of the humid northern subzone. It occurs also in the high limestone mountains of the middle and southern zones, and there it is usually a relict soil. A very interesting variety is the Terra fusca in the Serranía de Ronda in southern Spain under the famous relict forests of *Abies pinsapo*, the Spanish fir.

Humid Terra rossa is a decalcified soil with rotlehm fabric that occurs in the limestone areas of the northern subzone. Although quite common, it is here a relict soil, since the present environmental conditions do not permit the formation of a rotlehm fabric.

XEROMORPHOUS VARIETIES OF TERRA FUSCA AND TERRA ROSSA

In the middle and southern subzones of the Mediterranean, conditions favorable to the preservation of the humid varieties of Terra fusca and Terra rossa do not exist at the present time. Here it has become so dry that the relicts cannot remain in equilibrium with their environment. Such transformation as is produced consists in a loosening of the former dense and coherent fabric, partly by flocculation and disruption of the fabric, in an accelerated erosion which is considerably worsened by the lack of a dense enough protecting vegetation. Hence only a part of the older soil cover is preserved; in many habitats the soil has been completely removed. Other significant transformations consist in a strong secondary enrichment with calcium carbonate which sometimes causes the whole soil matrix to be impregnated with recrystallized calcite crystallites. If most of the older humus layer has been removed by erosion, it is difficult for a new humus horizon to form. Therefore these varieties are very poor in humus.

The arid transformation varieties also differ from the humid forms of Terra fusca and Terra rossa in that they contain a very large amount of allochthonous minerals, primarily in the silt and fine sand fraction. Their provenience can generally be easily determined. The strong dust formation and wind action of the middle and southern subzones, which greatly increase with proximity to the desert regions, account for these admixtures.

It is evident that the recalcified and considerably loosened Terra fusca and Terra rossa varieties in the middle and southern subzones of today's Mediterranean are transformed relict soils. They developed in the warm, humid periods of the Tertiary and of the Pleistocene. At the present time only Xero-Rendzinas can be formed in this region.

MAN'S INFLUENCE ON MEDITERRANEAN SOIL DEVELOPMENT

It is evident that mediterranean soils tend to an increase in xeromorphism at the present time, just as in the subtropics and the alternating dry and humid tropics. This is partly due to climatic changes. Those of us who have lived for a time in the Mediterranean area know that a few slightly more humid years are sufficient to produce striking changes in the vegetation and in the soil life. A succession of somewhat warmer and rainier winters can produce such changes.

The influence of man is also very crucial. The great demands for timber for shipbuilding in Italy, Spain and Greece, added to the always pressing need for firewood, have destroyed the woods of these countries over the years. The conversion of more land to agriculture and pasturage has further denuded the forest areas or reduced the stands of young timber. Other consequences are increased evaporation as more land area is exposed to the hot sun of the Mediterranean summer and the influence of drying winds. In other words, the xeromorphism of the soil cover has been enormously aggravated by man.

How can the impoverishment of soil and vegetation be at least partly reversed in the Mediterranean region? Some of the solutions lie in a thoroughly planned reforestation program for the wastelands to create protecting woods and a general increase in forest areas, in the adoption of good forestry practices and in the protection of the river systems, the construction of dams, an increase in irrigated areas and the development of fruit groves—in general, a well-organized erosion control and land conservation program.

XXIII

Soil on Loess in Different Parts of the Globe

Loess is a highly suitable parent material for micromorphological investigations undertaken in the course of soil development studies. Raw loess clearly reflects the smallest changes in environmental conditions, and hence the morphology of loess formations is enormously varied. These conditions suggest rich possibilities for future soil diagnostic studies. Indeed, micromorphological analysis of both loess soils and the soils of the tropics may be regarded as indispensable to a thorough understanding of the genesis of soils. Furthermore, loess is not just a parent material; it is also an independent soil formation, for it distinctly shows the influence of environmental conditions imposed on it during the time of its deposition.

ARCTIC RÅMARK

Loess as a soil represents a relict of an arctic Råmark, the terrestric raw soil of the arctic desert (Kubiëna, 1956c). This is proved by the weakly weathered state of the primary minerals, particularly the feldspars, augites and biotites, as well as by the mode of distribution of the calcium carbonate in the soil fabric and profile. Given a soil with high capillarity like loess in a hot, dry climate, accumulations of lime and lime crusts will form on the soil surface and on the walls of the soil cracks. This can be observed very clearly in the dense, fine marly sands of the dry areas of the islands of Tenerife and Grand Canary. The formation of loess in a cold, dry climate is also determined by the fact that the downward movement of calcium carbonate is slow; and this is true with all Braunerde and Lessivé varieties on loess. In calcareous raw soils in wet tundra regions, the undersides of the surface stones frequently exhibit accumulations of calcium carbonate. These lime accumulations are not produced by capillary rise; they represent travertine layers crystallized from the calcareous soil solution which reached the soil surface as a result of stagnation.

Raw loess as a soil formation of the cold desert was also recognized by Ložek (1965), largely on the basis of comparative investigations of the mollusk fauna. In his opinion, the mollusk shells are characteristic of raw loess, where the substrate is in a primitive form with little vegetative cover. Since they are different from the forms in all other glacial and interglacial communities, they must have developed under conditions peculiar to raw loess habitats. Furthermore, the special fabric in raw loess seems to be produced by special transformation processes comparable to those which take place as a result of the pulverization of soil layers in the dry desert (see Plate 9D, 9E, 9F). Ložek called the totality of these processes "loessification."

A true Loess Råmark does not occur as a recent soil formation. Thus the existing arctic raw soil layers are relict or fossil formations. The development of all loess soils is polygenetic, even those that display simple development sequences. To simplify our discussion of soil development in the temperate zone, the present raw loess will be designated as a Loess Syrosem. There are biological and morphological differences between it and Loess Råmark.

WET RAW SOILS

In the arctic plains of northern Alaska, Greenland and Spitsbergen, a distinctive raw soil is formed under semiterrestric conditions at the present time. Its micromorphology is characterized by conspicuous mud coatings on mineral grains and small aggregates. These coatings are dark to medium brown when investigated with transmitted light (see Plate 5B), but often look red, orange or ocher under incident light. They are produced because the sedimentation of the sand has taken place in muddy water. The heavy water movement that takes place on the soil surface as great masses of snow melt above the frozen soil in early summer increases the muddiness of the soil water considerably. The mud coatings rarely consist of an accumulation of humus substances (they have this appearance when examined with transmitted light at low magnification). In the majority of cases, they are pure clay and silt deposits that represent remnants of former non-arctic paleosoils. They are not produced by recent weathering of the primary minerals (these remain practically undecomposed). The red and orange colors are attributable to pseudogleyzation in the interior of the coatings. Thus the formation of the wet arctic raw soils is a polygenetic process.

A raw loess could never develop the micromorphology of the wet raw soils. This provides further reason to believe that the Loess Råmark existed as a soil in the dry polar desert.

TUNDRA RANKER AND TUNDRA RENDZINA

The term "tundra" refers to treeless areas with advanced polar vegetation and soil formation outside the desert areas. The lower layers of tundra soil may contain ground ice, and, as a consequence, stagnant soil water will appear in its surface layer during the summer season. On the other hand, tundra soil may be completely free of ground ice or stagnant water in rocky areas *(fiälls)* or in hilly country where the parent material is loose and easily dried out.

Therefore two groups of tundra soils must be distinguished: the terrestric and the semiterrestric. In the terrestric group Tundra Ranker and Tundra Rendzina are the most common varieties. Both are AC soils, the first occurs on silicate rocks, the second on limestone. If the parent material contains silicates as well as calcite or dolomite, the soil variety is called Tundra Pararendzina.

TUNDRA PARARENDZINA ON LOESS

There is no doubt that a well-developed Tundra Pararendzina on loess was a common formation during the glacial periods; it would have differed clearly from the Chernozems and Wet Chernozems (Pseudo-Chernozems, black prairie soils of the prairies of the temperate zones). This can be concluded from the nature of the many Tundra Rankers and Tundra Rendzinas of recent origin in the arctic. However, Tundra Pararendzina is very rare among the fossil soil formations of the Pleistocene. It is possible that it does exist among isolated buried humus remnants that have not yet been investigated micromorphologically. Their rareness may be attributable to the fact that an ABC soil develops much more rapidly on loess than on other parent materials.

TUNDRA ANMOOR ON LOESS

Dystrophic Tundra Anmoor is a common recent soil in the wet tundra. It is, to a great extent, developed by decomposition and humification of Carr Peats. Fossil Anmoor on loess has been found repeatedly in loess profiles. But these formations, combined with lessivé fabric and pseudogleyzation, evidently developed in a warmer climate.

THE PARARENDZINA OF THE TEMPERATE ZONE

This soil type represents the normal transition phase of soil development between the raw soil (Syrosem) and Braunerde or Parabraunerde. It occurs in the Braunerde region and in profile resembles a shallow Chernozem. Typically, it has a limited distribution. Occurring in a zone subject to intense chemical weathering, it is short-lived and is rapidly transformed into an ABC soil. It always develops as a Mull-Rendzina.

BRAUNERDE

Braunerde is the most common soil formation to develop monogenetically in the humid temperate subzone on silicate rocks, particularly if they are rich in bases and iron compounds. Braunerde on loess development is favored in the semicontinental section of the humid temperate subzone, whereas in the oceanic section Braunerde-Lessivé (leached Parabraunerde) development is greatly favored instead.

Braunerde on loess exhibits a clear flocculation of all substances of the clay fraction; the flocculated state of the clay in association with the amorphous iron hydroxide is easily seen. The scurfy, crumb-shaped aggregates are combined with each other into a slightly coherent sponge fabric which has high water stability. The humus form is mull; the profile normally has A(B)CaC horizons in which the calcium horizon generally is well developed (see Plate 6C, 6D and 6E).

ARCTIC BRAUNERDE

The arctic formation corresponding to the Braunerde of the temperate zone is the arctic Braunerde or the arctic Brown of Tedrow and Hill (1955). Like the alpine Sod Braunerde (see Plate 6B), it is characterized by an A(B)C profile, but the thickness of the horizons is greatly reduced, as is the whole profile. In regard to its micromorphology, the flocculation of the amorphous iron hydroxide and the other constituents of the fabric plasma is most characterizing; a loose sponge fabric with high pore space content is formed, as in alpine Råmark (see Plate 6B). Arctic Braunerde does not exist as a recent soil formation on loess. But it was found repeatedly as a fossil soil (Lieberoth, 1962).

BRAUNLEHM

A(B)C soils with braunlehm fabric are found not only in humid tropical environments but also in the subtropics, together with earthy Braunlehm. During the interglacial periods, particularly in low latitudes with warm and humid climatic conditions, there must have existed a transformation soil between the raw loess soils and the Braunlehm. This can be concluded from the finding of braunlehm-rich loess soils and Braunlehm-Lessivés on North Island of New Zealand. In the temperate zone, most of the braunlehm-rich soils on loess belong to the Braunerde-Lessivé group, and their genesis is different.

THE LESSIVÉ GROUP

A Lessivé is characterized by a compound micromorphology consisting of the fabric of a groundmass and the fabric of filling materials or the space wall coatings of its conducting channels or *Leitbahnen* (such as former root channels, earthworm tubes, shrinkage cracks). This separation is always produced by the presence of both a stable soil mass and a soil material with high mobility. The lessivé fabric is best developed in the B horizons, and sometimes also in the B/C horizon, of the soil profile.

According to the different combinations of fabrics in these two-phased, in some cases even three-phased, soils, different varieties of Lessivés can be distinguished. By comparative micromorphological investigations it can be demonstrated that there exist not just one form of Lessivé but at least five subtypes: Braunerde-Lessivé, earthy Braunlehm-Lessivé, Braunlehm-Lessivé, Rotlehm-Lessivé (see Plates 10 and 14), and Pseudo-gley-Lessivé.

BRAUNERDE-LESSIVÉ

The subtype Braunerde-Lessivé (leached Parabraunerde, Fahlerde) has a braunerde fabric in its groundmass and a braunlehm fabric in its conducting channels (*Leitbahnen*). On loess, it is a common soil with wide distribution. The designation "Parabraunerde" (Mückenhausen) was adopted by reason of its braunerde-like appearance and its occurrence in the Braunerde region. The best-developed Braunerde-Lessivé, however, is found in the oceanic section of the humid temperate subzone, where Braunerde on loess (never on basic silicate rocks) is very rare. In the Braunerde-Lessivé on loess of the oceanic section the amount of braunlehm plasma is greater, and it is much more peptized and movable; therefore, the lessivé fabric is always clearly visible to the naked eye. The plasma-filled conducting grooves may measure one to two centimeters in diameter in some cases. The best recent and Pleistocene Lessivé profiles are found in Belgium, the Netherlands, northern France and parts of West Germany (see Plate 14 and Tables 5 and 6). In the semicontinental section, the lessivé character of the soils is less apparent to the naked eye, the amount of braunlehm plasma being much lower and the plasma matrix less peptized and movable. In the steppes of the dry, winter-cold temperate subzone, Braunerde-Lessivés may occur as relict horizons on the surface and as fossil soils in the interior of the loess profiles. Usually the plasma-filled conducting grooves can be seen only under a microscope. The sparse Braunlehm remnants are irreversibly shrunk to almost horny masses and do not wilt when wetted. They are water stable and immovable (see Plate 14F). The fabric of the groundmass of the soil is still a braunerde fabric.

PSEUDOGLEY

Pseudogleys occur in all climatic zones in soil horizons subject to strong chemical weathering and temporary water stagnation. In Pseudogleys of the temperate zone, the mottling changes from white or light ocher to ocher; in the tropics, subtropics or Mediterranean countries, the iron-stained fabric parts usually change to orange or red. In all Pseudogleys the reduction rate is low, and the ferric hydroxides change during the pseudogleyzation from peptized amorphous to crystallized forms, mainly to aggregates of goethite, in some cases lepidocrocite. This is also true of tropical and subtropical Pseudogleys. Why the latter are more reddish is not yet known.

PSEUDOGLEY-LESSIVE

If the soil mass of such a tropical or subtropical Pseudogley shrinks because of dryness, a system of cracks is produced. After a new wetting by percolating water, iron-stained clay or silty clay moves into the cracks and these may be filled out entirely by the highly peptized plasma masses. The color of most of these filling substances is bright orange to red (see Plate 15E and 15F).

STAGNOGLEY

True gley formation can be produced by more or less complete reduction and secondary precipitation of partly or thoroughly crystalline ferric hydroxides under the influence of groundwater or a large amount of stagnant rainwater in the surface layer of the soil. The persistence of stagnant rainwater leads to the formation of Stagnogley (Vogel). The first Stagnogleys to be studied and described were those soil formations on loess in West Germany. They are typical for some forest areas on loess where the surface water always stagnates, where there is considerable reduction and where there is raw humus and dystrophic anmoor formation (see Plate 11A).

PSEUDO-CHERNOZEM

Pseudo-Chernozem (Wet Chernozem, black prairie soil) has been described repeatedly as a degraded Chernozem; however, transition forms to anmoor-like soils or even to Anmoors have been mentioned in connection with it. The difference between Pseudo-Chernozem and true Chernozem and its degraded forms is clear from their micromorphology.

To the naked eye the profile of a Pseudo-Chernozem resembles a degraded Chernozem with an A horizon 50 to 80 cm. (20 to 32 inches) thick which grades into a thin, seam-like or fully developed (B) horizon and continues into a Ca horizon. The Pseudo-Chernozem soils may represent transition forms to Parabraunerde as well as to Lessivé or even to Pseudogley. In all cases they are strongly weathered and, in contrast to the true Chernozem, they are characterized by a dense soil structure not only in the (B) horizon but also throughout the A horizon (see Plate 5E and 5F). The soil mass of the humus layer is blackish gray to black and is sticky when wet; it silts up easily and is highly erodible. In a dry state, it sometimes hardens to almost stone-like coherency; at the same time it may break down into angular fragments.

True Chernozem is a terrestric mull soil characterized by slow chemical weathering; it develops directly from the raw loess in a continuously growing humus layer. The degraded Chernozem is produced by increased chemical weathering under Braunerde conditions, and it forms a brim-like (B) horizon with a micromorphology similar to that of a Braunerde.

Pseudo-Chernozem is a tirsoid formation with an almost semiterrestric humus form. It is therefore found primarily on the lower parts of slopes or in depressions or poorly drained plains. It develops on highly weathered substrates, mostly on soil sediments. It is the tirs of the temperate zones, of regions with wet winters and a comparatively dry summer climate. It occurs primarily in the transition areas between the steppe and the humid regions: in the north-south belt of the earlier tall-grass steppe of the Middle West of the United States, and in the humid sections of the Polish, Ukrainian and East German plains.

Like every tirsoid soil, Pseudo-Chernozem develops on soil horizons or soil sediments (E/B horizons) of low permeability where water remains stagnant; the initial more or less anmoor-like humus formations are gradually transformed into tirsoid forms or even into mull.

During the spring (or wet winter seasons), the surface layers even in advanced development phases may become muddy and therefore display no aggregate formation. Aggregates may be found in the late spring, but since they have little water stability, they are completely destroyed after heavy rains. A striking breakdown of the soil structure may be observed on the soil surface, whereby a sedimentation of the eroded soil material, separated according to the different grain sizes, is produced by the runoff of the rainwater. Whitish streaks of completely washed sand and silt deposits alternate with blackish clay accumulations whose color is derived from humus substances. By contrast, true Chernozem has highly stable aggregates, and therefore it is highly permeable with great moisture capacity as soon as the soil thaws out completely in spring.

Micromorphological investigation may reveal stains or channel fillings of braunlehm plasma even in the A horizon of Pseudo-Chernozem. This plasma has an intense double refraction by particle arrangement. By comparison with such plasma in other soil formations, it is more enriched with humus substances and has a grayish tint when seen under transmitted light (Plate 5F). So much color can be added by the humus substances as to make the fluidal plasma almost black. During long wet periods the blackish, very mobile plasmoidic masses may move deep into the B horizon through open shrinkage cracks, earthworm tubes and root channels; such a movement never takes place in normal Chernozems.

CHERNOZEM

Genuine raw loess can be regarded as the most favorable parent material for the formation of Chernozem. In contrast to Pseudo-Chernozem, Chernozem has a fabric that is not dense; rather it is rich in aggregates (almost entirely earthworm casts) which may combine into a highly water-stable sponge fabric. The pore spaces are completely free of fluidal plasma (Plate 4F). The humus form is a well-developed terrestric mull; complete destruction of the soil structure and a redeposition of its separated constituents upon the soil surface do not take place.

GENESIS OF THE LESSIVÉ GROUP

Of the several loess soils, Braunerde-Lessivé, Pseudo-Chernozem, Pseudogley and the Stagnogley are, in essence, formations of a braunerde climate. Lessivé and to some extent even Pseudo-Chernozem indicate this by developing a braunerde fabric in their groundmass; only in Pseudogley and Stagnogley does the fabric become lost as a result of stagnating rainwater. Important for understanding the complicated polygenesis of Lessivé is the fact that, with adequate environmental conditions, the braunlehm fabric of the fluidal plasma in the conducting channels can be transformed into braunerde fabric, as shown in thin section in Plate 7C and 7D. This is produced in habitats with strong climatic influence, particularly cold winter seasons. The essential feature of a Braunerde-Lessivé (leached Parabraunerde), however, is the presence of an unchanged braunlehm plasma in the conducting channels (see Plate 14D).

Braunlehm fabric, according to numerous comparative investigations, is the product of a humid tropical or subtropical climate. Braunlehm fabric does not develop on crystalline silicate rocks in the temperate zone. But, excluding the subject of fossil soils, it has been found in the form of the braunlehm inclusions in a braunerde groundmass of a Lessivé on loess or other periglacial deposits. Therefore, the origin of the braunlehm plasma is the key to understanding the Lessivé formation.

We may recall a case of polygenetic development in the temperate zone in which braunlehm plasma was present in the limestone parent material in the form of remnants of former soils that had been produced by a humid tropical or subtropical climate of the geological past (see Chapter V). Their preservation had been much favored by their inclusion in the limestone. Then, as the calcium carbonate of the limestone dissolved, these fossil Braunlehm remnants accumulated in the form of sediments in depressions on lower slopes and slope steps, and new soil layers of braunlehm fabric were formed in the midst of a temperate region (this process is known as restitution). The genesis of this Terra fusca in the temperate zone can be easily traced by comparative micromorphological investigations of the soils and rocks involved and of the solution residues obtained by dissolving limestone samples in citric acid.

This history of Terra fusca development suggests a polygenetic explanation for the development of the Lessivé of the temperate zone, particularly since Lessivé varieties are formed not only from loess but also from moraines and gravel deposits, notably those rich in limestone gravels (Kubiëna, 1956c). In other words, there are instances in which gravels of crystalline silicate rocks produce braunerde fabric and gravels of limestone rocks restore masses of preformed braunlehm fabric.

The soils so formed look very much like a Terra fusca, but when they are investigated in thin section, they clearly have the micromorphology of a Braunerde-Lessivé. Hence the braunlehm plasma in Lessivé has its origin in the process whereby a fossil Braunlehm is restored from remnants which become enclosed in a soil mass characterized by braunerde fabric that developed from the gravels of silicate rocks under the influence of a winter-cold temperate climate. Under humid conditions the braunlehm fabric is still mobile and is transformed into the fluidal plasma, whereas the flocculated and earthened braunerde plasma remains in place. This accounts for the separation of the immobile groundmass and the mobile braunlehm plasma in the form of fillings of the conducting channels.

ORIGIN OF THE BRAUNLEHM PLASMA IN LOESS

If the braunlehm plasma in loess and other glacial deposits is not formed by direct weathering under the influence of a temperate climate, what other origin can be considered?

We know that the advance of a glacier removes a great deal of the soil cover, leaving bare the rock surface underneath. The soils eroded by the alpine glaciers in the present era are Braunerde, Semipodzol and Podzol varieties. What kinds of soils were removed from their parent rocks by the very active glaciers of the Pleistocene? The preglacial soil cover consisted primarily of subtropical soils or, if formed in the early Tertiary, even of tropical soils. Both these soils and the Tertiary sediments, consisting mainly of erosion masses of terrestric soils, were mainly of Braunlehm character. Even at the present time, most of the sediment relicts are Braunlehm sediments, whereas contemporary Braunerde sediments are found only on the lowest river terraces.

Most preglacial soils and soil sediments removed by the advancing glaciers have tended to accumulate among unweathered rubbles and mechanically ground rock detritus in moraines. This is very evident in areas that contain many varieties of sedimentary parent materials, as in southern and central England. The alternation of the old sediments can be recognized from the composition of glacial deposits in which they appear as characteristic constituents.

In dry polar deserts, the soil remnants in the moraines are pulverized and moved by wind erosion, to be deposited, together with the other constituents, in the form of loess. This raw loess represents the parent material for Loess Råmark. How did it happen that the smallest fraction of the former soils, the clay minerals, the iron hydroxides and the hydrated silicic acid, was deposited together with the much coarser material of the silt and the fine sand? The answer to this question lies in comparative micromorphological analysis of raw loess varieties.

NATURE OF THE BRAUNLEHM CONSTITUENTS IN RAW LOESS

The hypothesis that remnants of braunlehm plasma of former subtropical or tropical soils in the raw loess were reactivated after the decalcification of the upper loess layer has already been discussed above and in a previous publication (Kubiëna, 1956c). Under the very humid conditions of the interglacial periods, these remnants became mobile and moved into the conducting channels; this is why they now appear concentrated as channel fillings or pore space wall coatings typical for the lessivé fabric. The amount of fluidal plasma may even be increased by a possible peptizing effect of the braunlehm plasma, acting as a protective colloid upon other strongly weathered mineral substances.

This possibility was confirmed by micromorphological investigations of Braunerde and Lessivé varieties of New Zealand (Kubiëna, 1964). Near the east coast of South Island, a large part of the braunlehm plasma of these soils was derived from a Tertiary marl of the shelf region, and the cavities of this marl contained highly peptized Braunlehm fillings (see Plate 15A and 15B). When the sea regressed during the Pleistocene, the marl deposits became dry and open to loessification. In this process, the pulverized matter eroded by the wind consisted, not of a fine powder, but of small flaky fragments of the undisturbed braunlehm fabric. With intense desiccation, the fabric of the flakes can harden completely and become stable, as does a glauconite grain. These stabilized flakes will not become silted up, even under Braunerde conditions (see Plate 15A).

Identifying Braunlehm constituents in thin sections of the raw loess is greatly facilitated if incident light instead of transmitted light is used. Even if some slight flocculation of the amorphous hydroxide has taken place in the dense matrix (making it look dirty brown under transmitted light), the bright egg-yellow color of the Braunlehm shows up clearly under incident light.

The Braunlehm constituents may occur in the following forms:

(1) Angular or rounded fragments of flaky material similar to glauconite grains (see Plate 15C). These Braunlehm flakes consist of a dense bright-yellow, orange or red matrix. The morphology resembles that of flakes that sometimes occur in another raw soil, the Yerma of the subtropical or tropical dry desert. Plate 9D and 9E show an undisturbed

braunlehm fabric; in the upper part of the picture the morphology has been destroyed by pulverization under the influence of the dry desert. In Plate 9D, this transformation is seen at somewhat higher magnification. Here the loose surface layer of a soil crack consists of mineral grains which are coated with deposits of braunlehm plasma; angular ocherous fragments of a braunlehm matrix can also be seen.

(2) Flakes of braunlehm matrix which have lost their angular or rounded shape by flowing apart to form irregular excrescences (Plate 15D). These are produced by the influence of increased humidity upon the raw loess or polar Råmark.

(3) Braunlehm plasma in the form of deposits or irregular coatings on the surface of mineral grains (Plate 15D). These are mineral grains (mostly quartz) of a braunlehm fabric whose deposits or coatings were preserved during the erosion and transportation process. Similar sand grains can be found in dry desert soils (Plate 9D).

(4) Braunlehm plasma that has become mobile as a result of percolating water and has formed thin deposits of fluidal plasma in the intergranular spaces (Plate 15D).

(5) All forms of braunlehm plasma in the raw loess or in polar Råmark may be yellow, but they may also contain some reddish stains or be orange or bright red. Since all the transition stages of a development from yellow to red can be recognized and are well known to the micromorphologist as a result of comparative investigations of other soils, all can definitely be attributed to pseudogleyzation. This process, however, took place in the old Braunlehm profiles, before the glacial abrasion and the subsequent loessification and redeposition.

(6) In the form of fluidal plasma. With the increased humidity that developed with the climatic change from the glacial to the interglacial ages, the partly pseudogleyed Braunlehm remnants became mobile and could move into the conducting channels and accumulate there in the form of fluidal plasma. The pseudogleyed material sometimes turns solid orange in color after complete mixing (see Plate 15E). If mica-like clay minerals predominate in the matrix (Plate 15F), a striking double refraction of the fluidal plasma caused by particle arrangement can be observed. The red color is not produced by rubification (as previously thought) but is caused by pseudogleyzation of a variety typical for humid tropical, subtropical or partly mediterranean conditions.

The special nature of the Braunlehm remnants in raw loess or polar

Råmark described in paragraphs (1) and (3) of the preceding list requires some additional explanation.

Soils with a dense braunlehm fabric are produced in a humid tropical or subtropical climate. We know that this fabric is easily transformed if exposed to climates that are drastically different, especially if the winters are cold (see the description of brown earthening of Braunlehm, Plate 7). The decisive environmental factor in the formation of raw loess and Loess Råmark is an extremely dry desert climate. Severe drying of a braunlehm fabric leads to its stabilization, and this can make Braunlehm fragments almost unchangeable, even under the influence of a later humid Braunerde climate. That is why the Braunlehm flakes in the raw loess and in the Loess Råmark, that is, in soils of a dry polar desert, have almost the same character as the Braunlehm flakes of sediments deposited by the wind in dry tropical and subtropical deserts (see Plate 9D and Plate 15C).

GLOBAL DISTRIBUTION OF LOESS SOILS

Besides the descriptions of the soil types, subtypes and varieties on loess and their micromorphology, an over-all view of their global distribution by soil provinces is given in Table 6. Both present and past periods are taken into account. Thus the loess soils belonging to the arctic province developed in the Pleistocene period and are found today only in the form of relict or fossil soils. In the humid temperate province and in the steppe province the loess soils of the Pleistocene period were almost identical to those of the present time in Europe and in North and South America. Loess soils of the submediterranean, mediterranean and subtropical provinces are found in Europe and in North and South America, but they can also be found in New Zealand. In North Island of New Zealand there are Braunlehm-Lessivés and subtropical Pseudogleys. Towards the warmer north the Braunlehm content increases, towards the south and with the higher altitudes of South Island secondary brown earthening is very evident. A number of these soils have the character of a Braunerde-Lessivé (Parabraunerde).

The distinct increase in dryness in present environmental conditions in some regions, particularly the summer-dry areas of the Otago and the Canterbury districts, corresponds to the xeromorphism in the loess soils of southern Europe. With 330 mm. (13 inches) of rainfall, Alexandra, the center of Otago, has the lowest annual precipitation in New Zealand. After prolonged drying, soils of an originally warm and humid climate, such as the Braunlehm-Lessivés, may no longer lend themselves to transformation by brown earthening and may undergo an almost irreversible hardening (see the Braunlehm-Lessivé in Plate 10A and in Plate 14A).

In order to simplify the picture of the global distribution of the various loess soils the main types of the Burozem and Sierozem soils of the semi-desert district have been left out; so also have some soil varieties that represent a submediterranean variety of red Pseudogley in southern Europe, Hungary, Yugoslavia and southern France.

Table 6   Simplified Global Distribution of Recent and Fossil Loess Soils and Their Development Tendencies

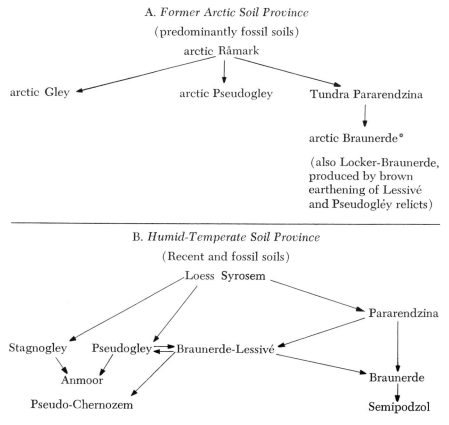

A. *Former Arctic Soil Province*

(predominantly fossil soils)

arctic Råmark

arctic Gley   arctic Pseudogley   Tundra Pararendzina

arctic Braunerde*

(also Locker-Braunerde, produced by brown earthening of Lessivé and Pseudogléy relicts)

B. *Humid-Temperate Soil Province*

(Recent and fossil soils)

Loess Syrosem

Pararendzina

Stagnogley   Pseudogley ⇌ Braunerde-Lessivé

Anmoor   Braunerde

Pseudo-Chernozem   Semipodzol

*Arctic Brown as described by J. C. F. Tedrow.

C. *Steppe Soil Province*

(recent and fossil soils)

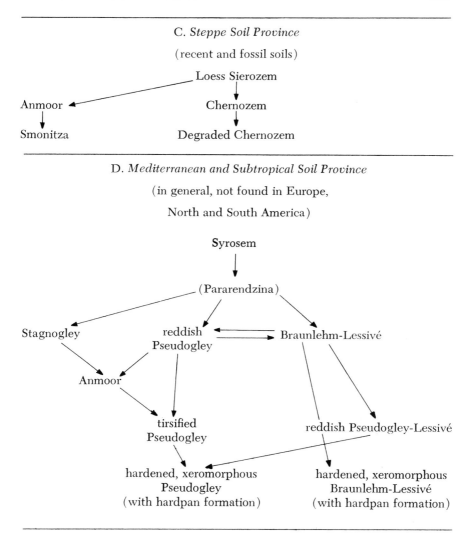

D. *Mediterranean and Subtropical Soil Province*

(in general, not found in Europe,

North and South America)

Appendix

The Climate Diagrams of Walter and Lieth

With the publication of the three-volume *Klimadiagramm-Weltatlas* of Heinrich Walter and Helmut Lieth of the Botanischen Institut der Landwirtschaftlichen Hochschule, Stuttgart-Hohenheim, in the 1960's, a valuable new tool for comparative investigations of climates in most regions of the earth has been made available. Drs. Walter and Lieth have designed and compiled a uniform set of climate diagrams for many of the major meteorological stations, and from these they have also constructed a series of climate cartograms delineating the borders of the climatic zones in each country or region according to the principal climatic types.

A climate diagram, by summarizing an immense number of climatological details in graph form, provides, so to speak, a portrait of a given meteorological station. Since all the diagrams are plotted to the same scale, in a standardized pattern, it is possible to compare moisture and temperature relationships as well as other environmental conditions in widely separated parts of the globe and hence to correlate climate with soil properties and type of vegetation. Details of the diagrams indicate the intensity of each climatic factor, and also seasonal rhythms. In Tables 7 and 8 selected diagrams have been grouped to illustrate the influence of continentality, altitude and latitude on climate and soils.

Immediately below is a key to the design of the Walter and Lieth climate diagrams as abstracted from the introductory explanations in Volume I of the *Klimadiagramm-Weltatlas.**

* Permission to reproduce climate diagrams from the *Klimadiagramm-Weltatlas* was kindly granted by the authors and by the publisher, VEB Gustav Fischer Verlag of Jena. The diagrams have been reduced approximately 10 per cent.

The Walter and Lieth Climate Diagrams

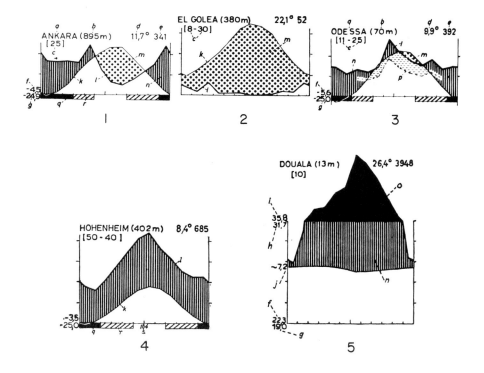

The Walter and Lieth Climate Diagrams

KEY   a—meteorological station (January is taken as the first month of
           the year in the Northern Hemisphere; July is the first month
           in the Southern Hemisphere. Read from left to right along the
           horizontal bar.)
           b—altitude (meters)
           c—number of years of observation [temperature-precipitation]
           d—mean annual temperature ° C. All temperatures are given in
           Centigrade.
           e—mean annual precipitation (millimeters)
           f—mean daily temperature of coldest month
           g—absolute minimum temperature recorded
           h—mean daily maximum temperature of hottest month
           i—absolute maximum temperature recorded
           j—mean daily range of temperature
           k—mean monthly temperature (thin line)
           l—mean monthly precipitation (millimeters) (thick line)
           Note: k and l are drawn in fixed proportion to 100 mm. of
           precipitation. Above this, the proportion is 1:10
           m—aridity (where precipitation line, l, falls below temperature
           line, k)
           n—humidity (where temperature line, k, falls below precipitation
           line, l)
           o—mean monthly precipitation greater than 100 mm.; plotted on
           a reduced vertical scale of 1:10
           p—special precipitation curve, plotted on a 1:3 scale (dashed
           line); a dashed line area is used to differentiate forest regions
           from the forest-steppe
           q—unfavorable (frost) season; mean monthly minimum tempera-
           ture below 0° C.
           r—absolute monthly minimum temperature below 0° C.
           s—mean number of frost-free days above 0° C., or above another
           limit of temperature

Table 7. Climate Diagrams of

A. Influence of Continentality in

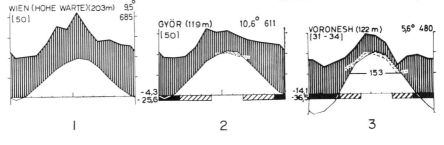

B. Influence of Altitude within the

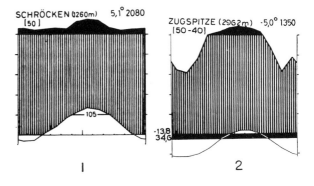

C. The Lowland Area (1) and the Cool Mountain
Forest Belt (2 and 3) of Equatorial South America

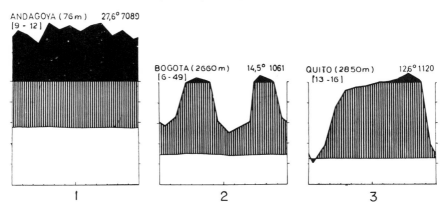

Plains and High Mountain Stations

the Austrian Alps and to the East

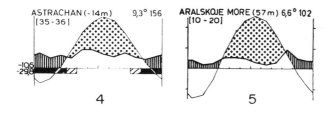

4   5

Austrian Alps from Northwest to Southeast

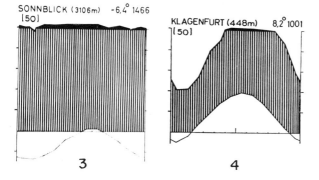

3   4

D. A Braunlehm Region (1) and a Rotlehm Region (2) of the Congo

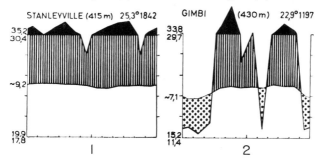

1   2

Table 8. Climate Diagrams of Europe

A. The Mediterranean Climate in Spain: Humid Northern Subzone (1 and
2); Semiarid Middle Subzone (3); Semihumid, Semiarid to Arid Subzone
(4, 5 and 6)

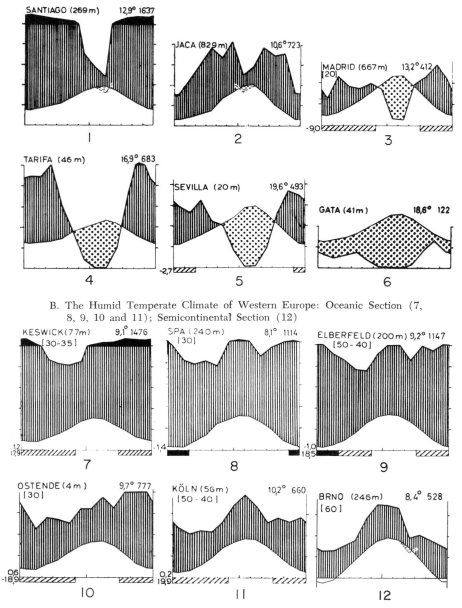

B. The Humid Temperate Climate of Western Europe: Oceanic Section (7,
8, 9, 10 and 11); Semicontinental Section (12)

Bibliography

Albareda Herrera, J. M. 1964. Die klimaxbildenden Ranker Spaniens, ihre Mikromorphologie und Genese. *In* A. Jongerius (ed.), *Soil Micromorphology*, Elsevier, Amsterdam, pp. 151–168.

Albareda Herrera, J. M., and Hoyos de Castro, A. 1955. *Edafología*. 2d ed. S.A.E.T.A., Madrid. 368 pp.

Altemüller, H.-J. 1956. Neue Möglichkeiten zur Herstellung von Bodendünnschliffen. *Z. f. Pflanzenernähr, Düng. u. Bodenk.*, 72:50–62.

―――――――. 1964. Die Anwendung des Phasenkontrastverfahrens bei der Untersuchung von Bodendünnschliffen. *In* A. Jongerius (ed.), *Soil Micromorphology*, Elsevier, Amsterdam, pp. 371–390.

―――――――. 1966. Die morphologische Untersuchung des Bodengefüges. *In* Karl Scharrer and Hans Linser (eds.), *Handbuch der Pflanzenernährung und Düngung*, Springer, Vienna; Vol. II, First Half, *Boden und Düngemittel*, pp. 230–263.

Altemüller, H.-J., and Frese, H. 1962. *Arbeiten aus dem Gebiet der Mikromorphologie des Bodens. Beitr. Erst Int. Arbeitstagung für Mikromorphologie des Bodens* (Braunschweig-Völkenrode). Verlag Chemie, Weinheim. 252 pp.

Babel, Ulrich. 1964. Dünnschnittuntersuchungen über den Abbau lignifizierter Gewebe im Boden. *In* A. Jongerius (ed.), *Soil Micromorphology*, Elsevier, Amsterdam, pp. 15–22.

―――――――. 1965. Humuschemische Untersuchung eines Buchen-Rothumus mittels mikroskopischer Methoden. *Mitt. Ver. Forstl. Standortsk. Forstpflanzenzücht.*, 15:33–38.

Beckmann, Walter. 1967. Bodenstruktur und Wasserhaushalt im Boden, dargestellt an Böden aus dem Südschwarzwald. *In* W. L. Kubiëna, et al., *Die mikromorphometrische Bodenanalyse*, Enke, Stuttgart, pp. 102–118.

Besoaín, M., and García Vicente, J. 1962. Clay Mineralogy of Some Volcanic Ash Soils of Chile. *Neues Jahrb. Mineral. Abhandl.*, 98:349–366.

Brewer, Roy. 1964. *Fabric and Mineral Analysis of Soils*. John Wiley and Sons, New York. 470 pp.

Buch, M.-W. von. 1967. Die Verwendung von Präzisions-Flächenschleifmaschinen bei der serienmässigen Herstellung von Bodendünnschliffen. *In* W. L. Kubiëna, et al., *Die mikromorphometrische Bodenanalyse*, Enke, Stuttgart, pp. 19–29.

211

Buchanan, F. 1807. *A Journey from Madras Through the Countries of Mysore, Canara and Malabar (1800–1801)*. London. Vol. II, pp. 436–460.

Büdel, Julius. 1955. Reliefgenerationen und Plio-Pleistozäner Klimawandel im Hoggar-Gebirge (Zentral Sahara). *Erdkunde*, 9:100–115.

————. 1960. Die Gliederung der Würmkaltzeit. *Würzburger Geogr. Arb.*, 8.

Commonwealth Bureau of Soils. 1965. *Bibliography on Micropedology, 1955–1965*. Harpenden, England.

Day, A. L., and Shepherd, E. S. 1913. Water and Volcanic Activity. *Bull. Geol. Soc. Amer.*, 24.

Dokuchaev, V. V. 1879. The Russian Chernozem. (In Russian.) St. Petersburg.

Dumont, J. 1913. *Etude sur le sol. I. Agrochimie*. Paris.

Esenwein, P. 1929. Die Petrographie der Azoren. *Z. f. Vulkanol.*, 12:108.

Fink, J. 1956. Zur Korrelation der Terrassen und Lösse in Österreich. *Eiszeitalter u. Gegenwart*, 7.

Franz, H. 1950. *Bodenzoologie als Grundlage der Bodenpflege, mit besonderer Berücksichtigung der Bodenfauna in den Ostalpen und im Donaubecken*. Akademie-Verlag, Berlin. 316 pp.

————. 1956. Drei klimabedingte Ranker-Subtypen Europas. *Trans. Sixth Int. Congr. Soil Sci.* (Paris), 5:135–141.

————. 1960. *Feldbodenkunde als Grundlage der Standortsbeurteilung und Bodenwirtschaft, mit besonderer Berücksichtigung der Arbeit im Gelände*. Fromme, Vienna. 583 pp.

Frei, Erwin. 1964. Micromorphology of Some Tropical Mountain Soils. *In* A. Jongerius (ed.), *Soil Micromorphology*, Elsevier, Amsterdam, pp. 307–311.

Friedel, H. 1934. Boden- und Vegetationsentwicklung am Pasterzenufer. *Carinthia* (Klagenfurt), 2:133.

Friedländer, Immanuel. 1929. Die Azoren. *Z. f. Vulkanol.*, 12:77.

Gallego. 1956. Personal communication.

Ganssen, R., and Hädrich, F. 1965. *Atlas zur Bodenkunde*. Bibliogr. Inst. Mannheim.

Geyger, Erika. 1967. Bodenstruktur und Entwicklung der Panama-Disease in Bananenpflanzungen. *In* W. L. Kubiëna, et al., *Die mikromorphometrische Bodenanalyse*, Enke, Stuttgart, pp. 135–162.

Glinka, K. D. 1914. *Die Typen der Bodenbildung, ihre Klassifikation und geographische Verbreitung*. Bornträger, Berlin. 365 pp.

Hamilton, R. 1964a. Microscopic Studies on Laterite Formation. *In* A. Jongerius (ed.), *Soil Micromorphology*, Elsevier, Amsterdam, pp. 269–276.

————. 1964b. A Short Note on Droplet Formation in Ironcrusts. *In* A. Jongerius (ed.), *Soil Micromorphology*, Elsevier, Amsterdam, pp. 277–278.

Hernández Pacheco, F. 1949. Las rañas de las sierras centrales de Estremadura. *Compt. Rend. Seizième Congr. Int. Géogr.* (Lisbon).

Hoyos de Castro, A., and Mata, A. 1959. *Alunitas del pico del Teide (Canarias)*. Barcelona.

Hoyos de Castro, A., and Pino, C. 1958. Estudio comparado de limos canarios y de Guinea Española. *Agrochimica*, 2:118–135.

Jaritz, G. 1966. Untersuchungen an fossilen Tertiärböden und vulkanogenen Edaphoiden des Westerwaldes. Inaugural dissertation, Bonn.

Jongerius, A. (ed.). 1964. *Soil Micromorphology*. Elsevier, Amsterdam. 540 pp.

Kress-Voltz, Marina. 1964. Gefüge- und Strukturuntersuchungen an vulkanogenen Edaphoiden. *In* A. Jongerius (ed.), *Soil Micromorphology*, Elsevier, Amsterdam, pp. 139–150.

_____. 1967. Mikromorphologische und mikromorphometrische Untersuchungen an vulkanischem Material und dessen praktische Bedeutung. *In* W. L. Kubiëna, et al., *Die mikromorphometrische Bodenanalyse*, Enke, Stuttgart, pp. 92–101.

Kubiëna, W. L. 1937. Verfahren zur Herstellung von Dünnschliffen von Böden un ungestörter Lagerung. *Zeiss Nachr.* 2(3):81–91.

_____. 1938. *Micropedology*. Collegiate Press, Ames, Iowa. 243 pp.

_____. 1942. Dünnschlifftechnik in der Bodenuntersuchung. *Forschungsdienst* (Berlin), Sonderheft 16.

_____. 1948. *Entwicklungslehre des Bodens*. Springer, Vienna. 215 pp. and plates.

_____. 1953a. *The Soils of Europe*. Thomas Murby, London. 317 pp. (In German: Enke, Stuttgart.)

_____. 1953b. Sur la reconnaissance de l'altération primaire et secondaire dans les sédiments désertiques. *Compt. Rend. Dixneuvième Congr. Int. Geol.* (Algiers).

_____. 1954a. Neue Beiträge zur Kenntnis des planetarischen und hypsometrischen Formenwandels der Böden Afrikas. *Stuttgarter Geogr. Stud. Lautensach Festschr.*, 69.

_____. 1954b. Micromorphology of Laterite Formation in Río Muni (Spanish Guinea). *Trans. Fifth Int. Congr. Soil Sci.* (Leopoldville), 4:77–84.

_____. 1954c. Sobre el método de la paleoedafología. *An. Edafol. Fisiol. Veg.* (Madrid), 12:523–543.

_____. 1955a. Reliktböden in Spanien. *Angew. Pflanzensoziol. Festschr. Aichinger* (Vienna), 1:214–224.

_____. 1955b. Über die Braunlehmrelikte des Atakor (Zentral Sahara). *Erdkunde*, 9:115–132.

_____. 1956a. Rubefizierung und Laterizierung, *Trans. Sixth Int. Congr. Soil Sci.* (Paris), Com. 5: E:247–249.

_____. 1956b. Materialien zur Geschichte der Bodenbildung auf den Westkanaren. *Rappt. Sixième Congr. Int. Sci. Sol* (Paris), 38:241–246.

_____. 1956c. Die rezenten und pleistozänen Böden auf Löss. *Eiszeitalter u. Gegenwart*, 7:102–112.

_____. 1958. Los suelos de los territorios españoles del Golfo de Guinea. *Arch. Inst. Estud. Afr.* (Madrid), 10(46):65–67.

_____. 1959. Prinzipien und Methodik der paläopedologischen Forschung im Dienste der Stratigraphie. *Z. Deut. Geol. Ges.*, 111(3):562–643.

_____. 1962. Die taxonomische Bedeutung der Art und Ausbildung von Eisenoxydhydratmineralien in Tropenböden. *Z. f. Pflanzenernähr, Düng. u. Bodenk.*, 98:205–213.

————. 1963. Paleosoils as Indicators of Paleoclimates. *In Changes of Climate, Proc. Symp. UNESCO-W.M.O.* (Rome), pp. 207–209.

————. 1964. Zur Mikromorphologie und Mikromorphogenese der Lössböden Neuseelands. *In* A. Jongerius (ed.), *Soil Micromorphology,* Elsevier, Amsterdam, pp. 211–235.

————. 1967. Einfluss des Bodens auf die Intensität der Ausbreitung und Entwicklung von Pflauzensuchen in den Tropen. *In Die mikromorphometrische Bodenanalyse,* Enke, Stuttgart, pp. 119–134.

Kubiëna, W. L., et al. 1967. *Die Mikromorphometrische Bodenanalyse.* Enke, Stuttgart. 196 pp.

Kühnelt, W. 1950. *Bodenbiologie mit besonderer Berücksichtigung der Tierwelt.* Vienna.

————. 1961. *Soil Biology with Special Reference to the Animal Kingdom.* London.

Kukla, Jiri. 1961. Lithologische Leithorizonte der tschechoslowakischen Lössprofile. *Vest. Ustredního Úst. Geol.,* 5:369–372.

Laatsch, W. 1957. *Dynamik der mitteleuropäischen Mineralböden.* Steinkopff, Dresden. 280 pp.

Lautensach, Hermann. 1952. *Der geographische Formenwandel.* Colloquium, Bonn.

Le Châtelier, H.-L. 1913. *La Cilice et les silicates.* Paris.

Lieberoth, Immo. 1962. Über den Einfluss der Ackerkultur auf die Bodenentwicklung in sächsischen Lössgebiet. *Albrecht-Thaer-Arch.,* 6:3–30.

Lozek, Vojen. 1964. The Relationship Between the Development of Soils and Faunas in the Warm Quaternary Phases. *Sborn. geol. Ved.,* 3:7–31.

————. 1965. Das Problem der Lössbildung und die Lössmullusken. *Eiszeitalter u. Gegenwart,* 16:61–75.

Marbut, C. F. 1927. A Scheme for Soil Classification. *Proc. First Int. Congr. Soil Sci.* (Washington, D.C.), IV.

Mella Lagos, A. 1958. Micromorphological Study of Some Volcanic Soils in Chile. (In Spanish.) *Agricultura Tei.* (Santiago), 18:166–184.

Merck, E. 1922. Kieselsäure. *Wiss., Abhandl.* 34.

Meyer, Brunk. 1960. Zeitmarken der Entwicklung mitteldeutscher Löss- und Kalksteinböden. *Seventh Int. Congr. Soil Sci.* (Madison, Wisconsin), 25: 177–183.

Miklaszewski, S. 1922. Contribution à la connaissance des sols nommés rendsinas. *Compt. Rend. Troisième Conf. Agropedol.* (Prague).

Mückenhausen, E. 1959. *Die wichtigsten Boden der Bundesrepublik Deutschland.* Kommentator, Frankfurt a. M. 146 pp.

————. 1967. Die Feinsubstanz vulkanogener Edaphoide. *An. Edafol. y Agrobiol.,* 26:1–4, pp. 703–715.

Müller, P. E. 1879. *Studies of the Natural Humus Forms.* (In Danish.) Copenhagen.

————. 1887. *Studien über die natürlichen Humusformen und deren Einwirkung auf Vegetation und Boden.* Springer, Berlin. 324 pp.

Oberdorfer, E. 1965. Pflanzensoziologische Studien auf Teneriffa und Gomera. *Beitr. Naturforsch. SW-Deutschl.*, 24:47–104.

Odén, S. 1912. Zur Kenntnis der Humussäure des Sphagnum-Torfes. *Ber. Deut. Chem. Ges.*, 35:651.

Oehme, R. 1935. Die Rañas: Eine spanische Schuttlandschaft. *Z. f. Geomorphol.*, IX(1):25–42.

————. 1942. Beiträge zur Morphologie des mittleren Estremadura. *Ber. d. Naturforsch. Ges. zu Freiburg i. Br.*, 38.

Pallmann, H. 1942. Grundzüge der Bodenbildung. *Schweiz. Landw.*, 22:1–4.

Pallmann, H., and Haffter, P. 1933. Pflanzensoziologische und bodenkundliche Untersuchungen im Oberengadin. *Ber. d. Schweiz.-Botpflanz. Ges.*, 42: 357–466.

Primavesi, A. 1961. Personal communication.

————. 1964. *Proc. Eighth Int. Congr. Soil Sci.* (Bucarest), Com. 4:123–125.

————. 1966. *Edafologia, Geo-Biologia, Nutrição Vegetal*, Orgão do Inst. do Solo e Cultura (Santa Maria, Brazil), 3.

Ramann, E. 1905. *Bodenkunde*. Berlin.

Scheffer, F., and Meyer, B. 1958. Ein bodenkundlicher Beitrag zur Grabung im Muschelkalkgebiet des Hünstollens bei Göttingen. *Göttinger Jahrb.*, pp. 3–7.

Scheffer, F., Welte, E., and Ludwig, F., 1957. Zur Frage der Eisenoxydhydrate im Boden. *Chem. der Erde* (Jena), pp. 51–64.

Schmidt-Lorenz, R. 1964. Zur Mikromorphologie der Eisen- und Aluminium-oxydanreicherung beim Tonmineralabbau in Lateriten Keralas und Ceylons. *In* A. Jongerius (ed.), *Soil Micromorphology*, Elsevier, Amsterdam, pp. 279–290.

————. 1967. Mikromorphologie und Mineralogie tropischer Pseudogleye, Gleye und lateritischer Fleckentone. *Arbeitsber. Deut. Forschungsgemeinsch.* (Ms.) (Hamburg-Reinbek.)

————. 1970. Die Böden der Tropen und Subtropen. *In* P. V. Blanckenburg and H. D. Cremer, *Handbuch der Landwirtschaft und Ernährung im den Entwicklungsländern*, Vol. II.

Smolíková, L. 1962. Zur Altersfrage der mitteleuropäischen Terrae calcis. *Eis zeitalter u. Gegenwart*, 13:157–177.

————. 1963. Stratigraphische Bedeutung der Terrae calcis-Böden. *Sborn. Geol. Ved. A.*, 1:101–126.

————. 1968. Mikromorphologie und Mikromorphometrie der pleistozänen Bodenkomplexe. *Rozpr. Ceskoslov. Akad. Ved.*, 78: 2:1–47.

Stebutt, A. 1930. *Lehrbuch der allgemeinen Bodenkunde*. Bornträger, Berlin, 518 pp.

Tedrow, J. C. F. 1962. Arctic Soils. *Proc. Permafrost Int. Conf. NAS-NRC*, p. 1287.

————. 1966. Polar Desert Soils. *Proc. Soil Sci. Soc. Amer.* 30:381–387.

Tedrow, J. C. F., and Cantlon, J. E. 1958a. Concept of Soil Formation and

Classification in Arctic Regions. *"Arctic," J. Arctic Inst. North Amer.,* 11:166–179.

Tedrow, J. C. F., Drew, J. V., and Douglas, L. A. 1958b. Major Genetic Soils of the Arctic Slope of Alaska. *J. Soil Sci.,* 9:34–35.

Tedrow, J. C. F., and Hill, D. E. 1955. Arctic Brown Soil. *Soil Sci.,* 80:265–275.

Troll, C. 1944. Strukturböden, Solifluktion und Frostklimate der Erde. *Geol. Rundschau.,* 34:718.

————. 1955. *Die dreidimensionale Landschaftsgliederung der Erde.* Bonn.

————. 1959. Die tropischen Gebirge, ihre dreidimensionale klimatische und pflanzengeographische Zonierung. *Bonner Geogr. Abhandl.,* 25. 93 pp.

Uhlig, Harold. 1966. Die Sierra Nevada de Santa Marta (Kolumbien). *Natur u. Mus.* (Frankfurt a. M.), 96 (2).

Vageler, P. 1930. *Grundriss der tropischen und subtropischen Bodenkunde, für Pflanzer und Studierende.* Verlagsgesellschaft für Ackerbau m.b.h., Berlin. 216 pp.

Vogel, F. Personal communication.

Walter, H. 1957a. Die Klimadiagramme der Waldsteppen und Steppengebiete in Osteuropa. *Stuttgarter Geogr. Stud. Lautensach Festschr.,* 69:253–262.

————. 1957b. Wie kann man den Klimatypus anschaulich darstellen? *Umschau,* 24:751–753.

Walter, H., and Lieth, H. 1960–1967. *Klimadiagramm-Weltatlas.* VEB Gustav Fischer, Jena.

Weidenbach, F. 1952. Gedanken zur Lössfrage. *Eiszeitalter u. Gegenwart,* 2.

Wiche, K. 1951. Personal communication.

Wohltmann, F. 1892. *Handbuch der tropischen Agrikultur.*

Wolff, F. von. 1914–1927. *Der Vulkanismus.* Vol. II.

Wyssotski, G. N. 1901. Das Illuvium und die Struktur der Steppenböden. *Potschwowjedjenie,* 3:137.

Zachariae, G. 1964. Welche Bedeutung haben Enchytraeen im Waldboden? *In* A. Jongerius (ed.), *Soil Micromorphology,* Elsevier, Amsterdam, pp. 57–68.

————. 1965. Spuren tierischer Tätigkeit im Boden des Buchenwaldes. *Beih. z. Forstwiss. Zentralbl.,* 20:1–68.

————. 1967. Personal communication.

COLOR PLATES

Plate 1.

Development of Fungi in the Raw Humus of a Spruce Forest with Iron Humus Podzol near Hamburg, West Germany

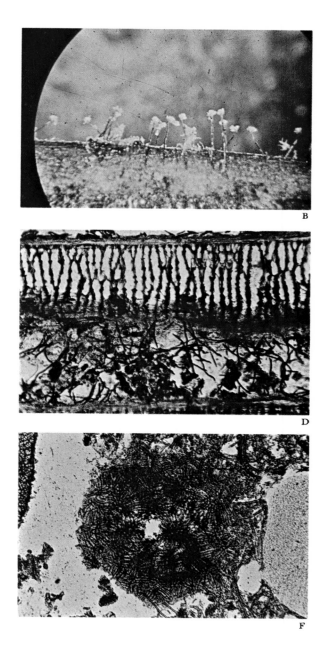

Plate 2.

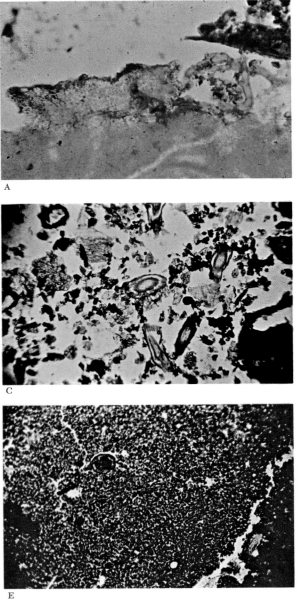

A

C

E

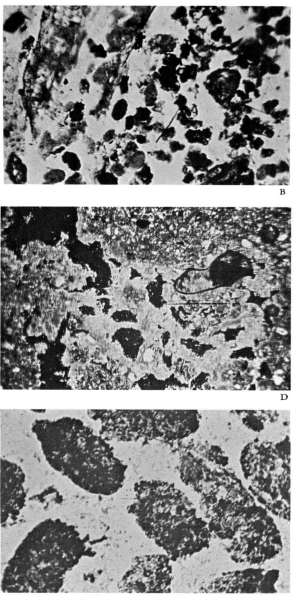

Plate   3.

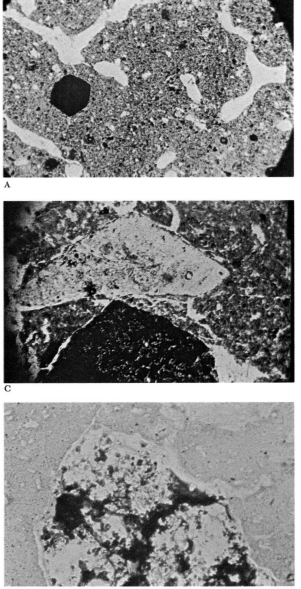

A

C

E

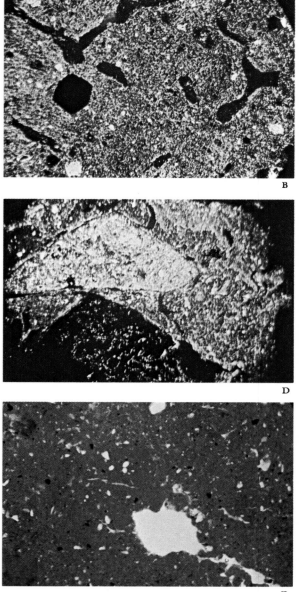

B

D

F

Plate 4.

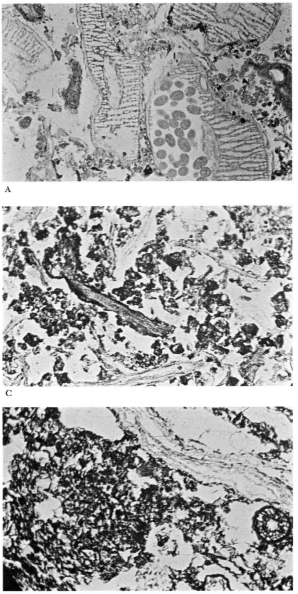

A

C

E

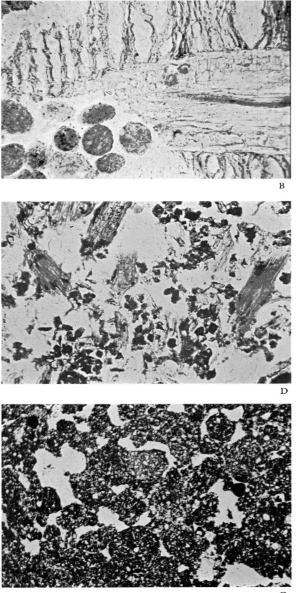

B

D

F

Plate 5.

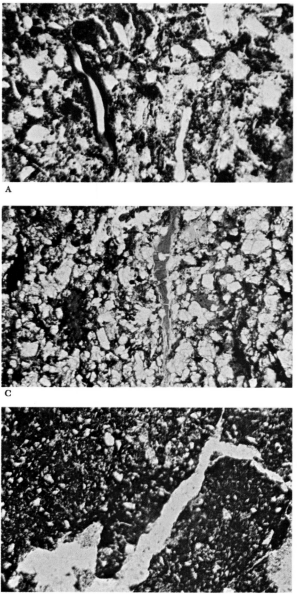

A

C

E

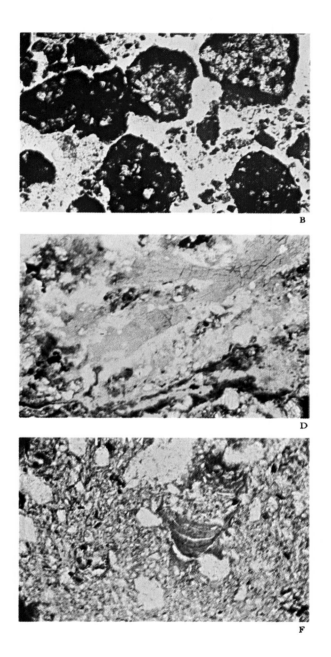

Plate 6.

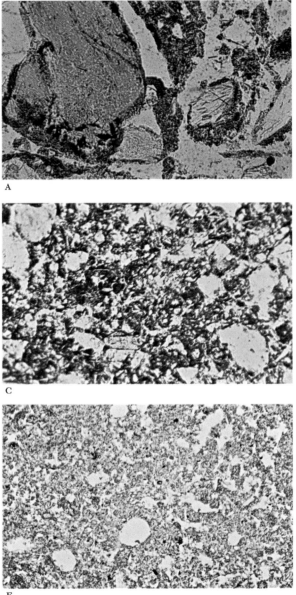

A

C

E

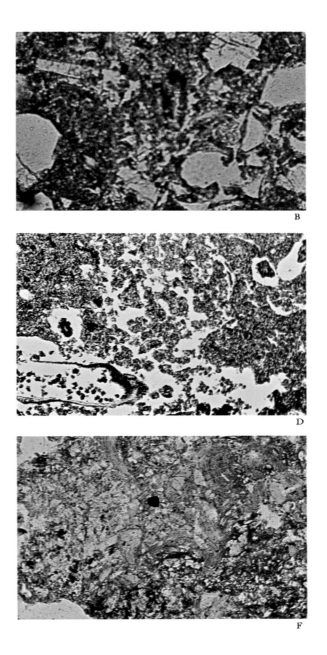

B

D

F

Plate 7.

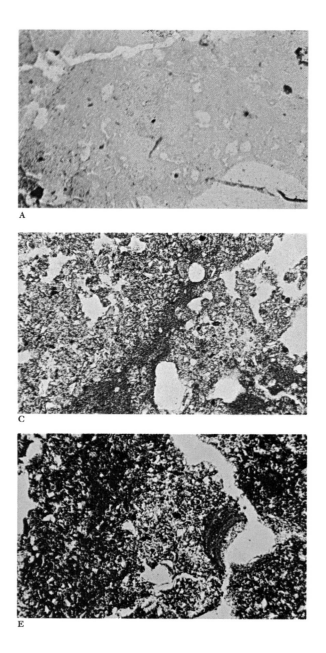

A

C

E

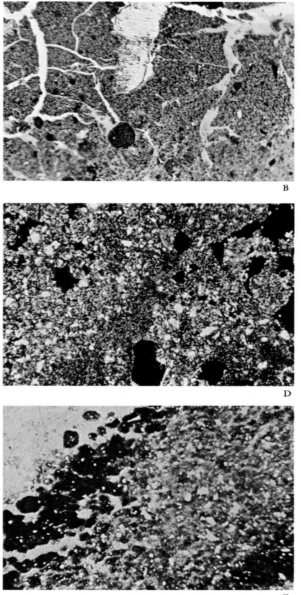

B

D

F

Plate   8.

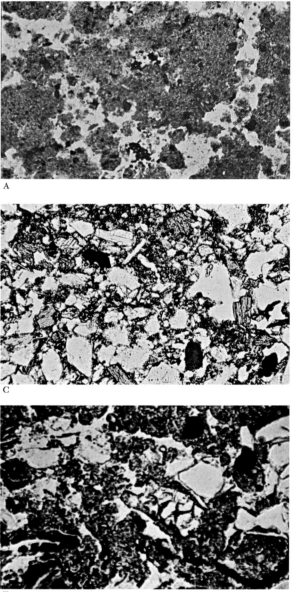

A

C

E

Earthy Braunlehm

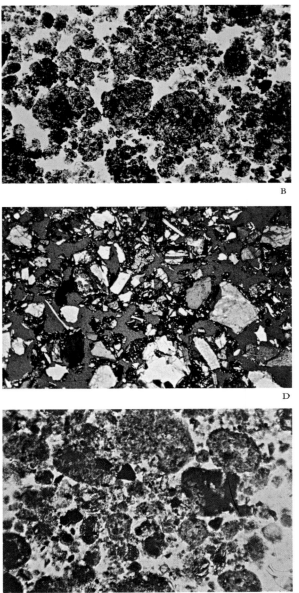

B

D

F

Plate 9.

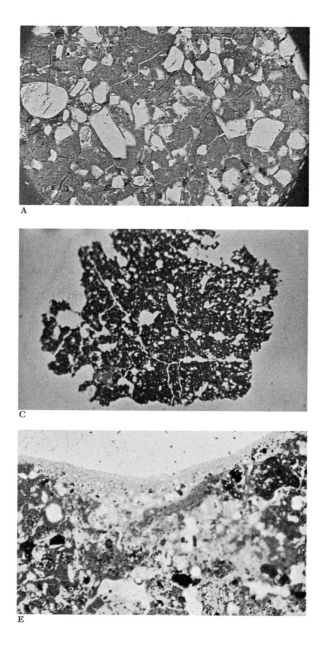

A

C

E

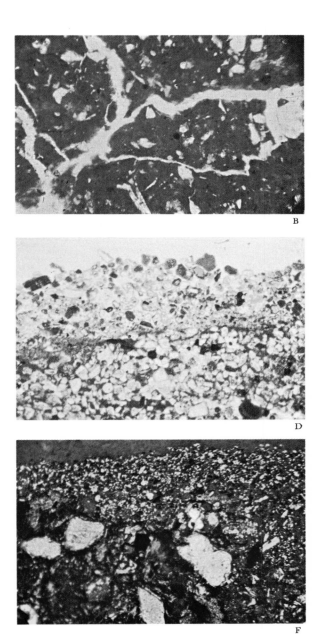

B

D

F

Plate 10.　　Lessivé Varieties

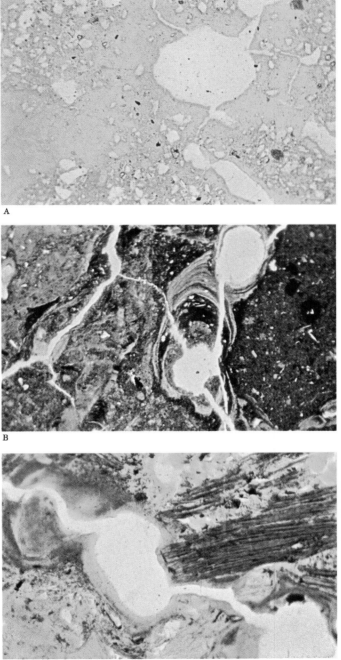

A

B

C

Plate 11.

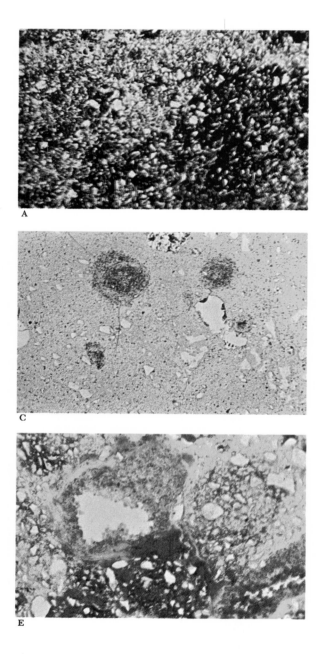

A

C

E

Pseudogleyzation

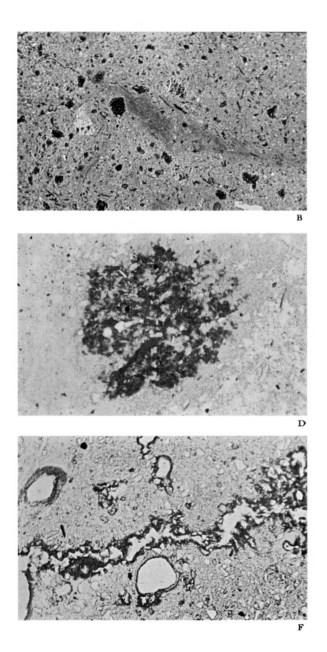

Plate 12.

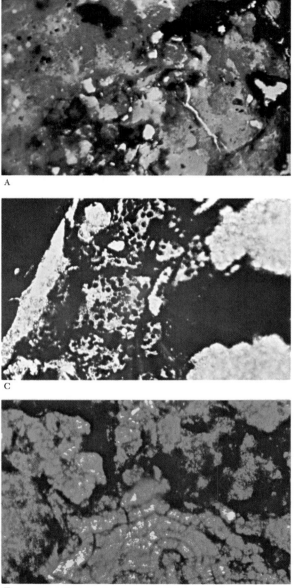

A

C

E

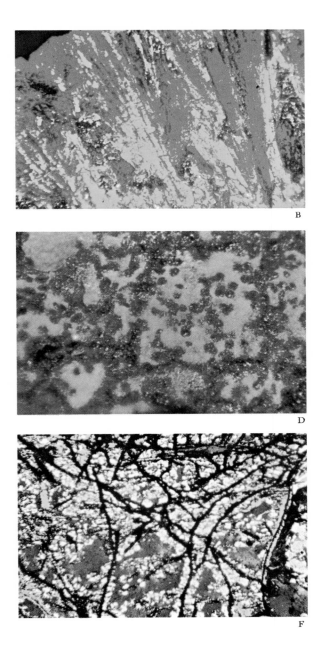

B

D

F

Plate 13. Vulcanogenic Edaphoids

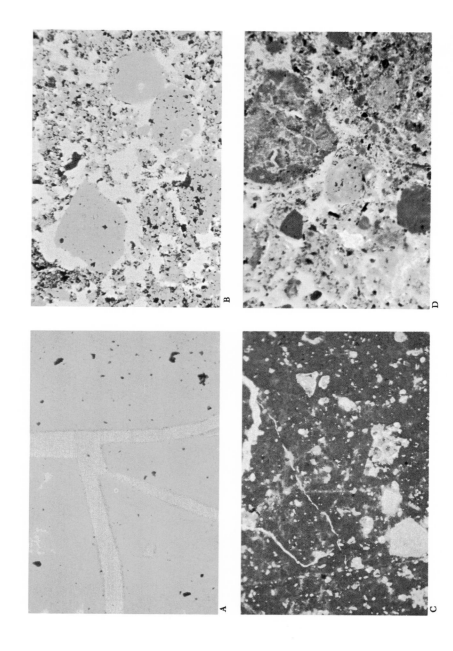

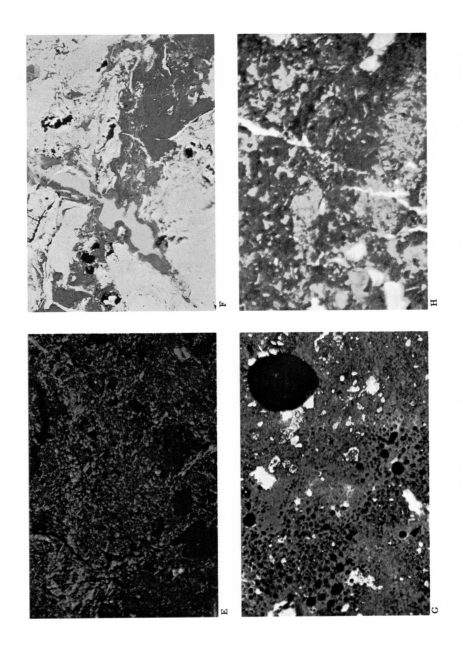

Plate 14.

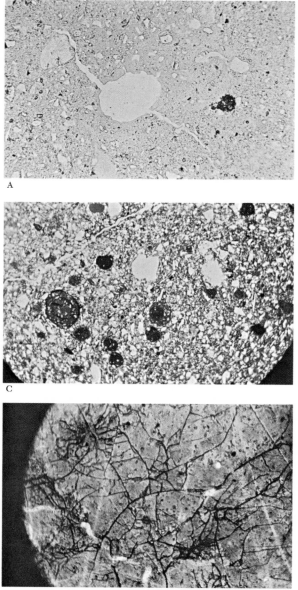

A

C

E

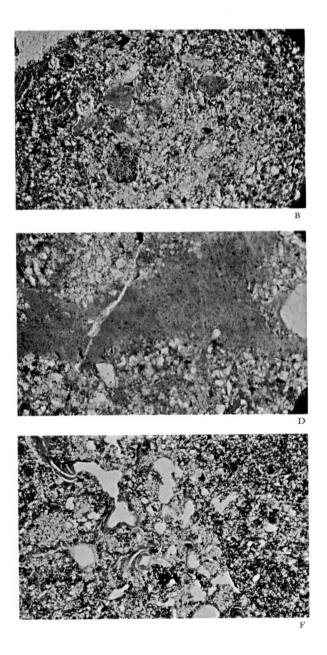

Plate 15.

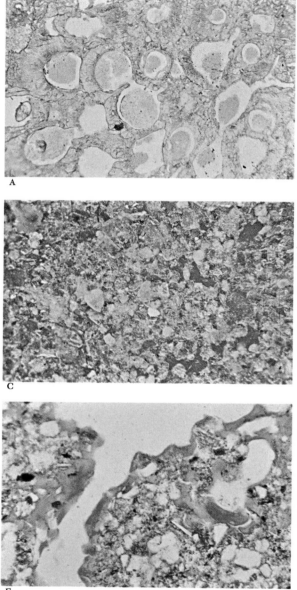

A

C

E

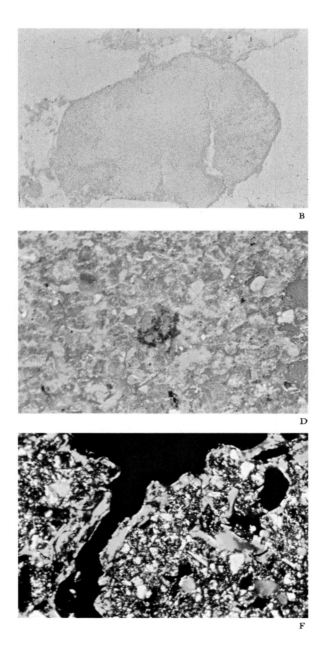

B

D

F

DESCRIPTIONS OF THE
COLOR PLATES

PLATE

Plate 1

DEVELOPMENT OF FUNGI IN THE RAW HUMUS OF A SPRUCE FOREST WITH IRON HUMUS PODZOL NEAR HAMBURG, WEST GERMANY

A.   Colony of yellow mycelium of *Oöspora sulphurea* in the A_H horizon. Direct microscopy of the unprepared soil with incident light. Within the colony are some plant roots; outside it are undecomposed plant residues and cylindrical blackish droppings of small arthropods.

B.   Unprepared spruce needle of the A_F horizon with fruiting bodies of *Hormodendron* species on its surface.

C.   Detail of the fruiting bodies of 1B seen with incident light. Note the brown stems and white conidia.

D.   A spruce needle of the A_F horizon containing the mycelium of *Hormodendron* species. Dark-brown mycelium and blackish droppings of horn mites are present in the feeding cavity of the lower part. In the upper part, the mycelium is growing in the interior of the cell walls and will develop through the epidermis of the needle to its surface, where, as seen in 1B and 1C, the fruiting bodies are formed.

E.   The development of a mycelium mantle of *Coenococcum graniforme* on spruce roots (half decomposed).

F.   Detail of the mycelium mantle of 1E showing some of the characteristics of the *Coenococcum* hyphae. They are dark brown and septated, and some are irregularly bent and swollen to various thicknesses.

Plate 2

HUMUS FORMATION ON LIMESTONE ROCKS

A.   Sequences of colonization on the surface of a limestone rock, southern Vienna Woods, Austria. Rhizoids of a moss cushion of *Grimmia orbicularis* (from right to left) have penetrated the surface crust of the rock underneath the alga layer of the lichen *Lecanora crassa*. The

219

Lecanora developed on an area formerly occupied by an endolithic lichen, *Verrucaria calciseda*. Both lichens, particularly the *Verrucaria*, have loosened the surface layer of the limestone rock in creating small cavities by solution weathering, and this has considerably favored colonization. The rhizoids of the moss, which here are only beginning the grow, rapidly develop into a dense network. 36×.

B.   The moder of a Proto-Rendzina on calcareous dolomite under a grass sod of *Sesleria varia*, southern Vienna Woods, Austria. The very loose humus formation consists of partly decomposed plant fragments, cylindrical droppings of horn mites and small amounts of rock débris. The reddish droppings (left), which are derived from horn mites feeding on the reddish plant fragments, are little humified. The blackish droppings, which are derived from more decomposed plant residues with a lower lignin content, are more humified but as yet contain no visible mineral constituents in their interior. The form of the droppings is well preserved, and they measure about 60μ in length. Small quantities of calcite (dolomite) grains are present, but only outside the droppings. Some have well-preserved crystallographic delimitation, for example, the light-gray grain in the center. The browning of the lignin-rich plant tissues is evidently produced by fungi. Their dark-brown mycelium is visible in the reddish-brown plant tissue. Note an isolated dark-brown fungus mycelium (cut off on both ends) in the center, below the rhombohedral calcite (dolomite) grain.

C.   An alpine Proto-Rendzina, Austrian limestone Alps. This primitive humus formation consists of little-decomposed plant fragments, of blackish droppings of small arthropods whose organic matter is fairly well humified, and of small limestone débris (fine and coarse sand size). The soil mass is completely loose; there is no aggregate formation. Mineral substance is present only in the form of clastic calcium carbonate, that is, in the form of mechanically weathered limestone. The humus form is rendzina-moder. Here, as in 2B, the crystallographic forms (rhombohedral fragments) are well preserved. Clastic calcium carbonate in upper Rendzina layers indicates humid Rendzina varieties, whereas recrystallized calcium carbonate in surface layers, reprecipitated from the soil solution, indicates Xero-Rendzinas.

D.   Recrystallized calcium carbonate in spaces of the A/Ca horizon of a calcareous mull formation. The crystals take the form of very small, rounded crystallites or small needles like the pseudomycelium crystals in Chernozem. They are formed by efflorescence from the porous space walls into the interior of the air-filled soil space. In dry soils this kind of recrystallization reaches into the upper humus horizon and to the surface of the soil.

E.   An alpine Pitch-Rendzina, Dachstein, Austrian limestone Alps. The humus mass consists almost entirely of cylindrical blackish droppings of small arthropods. The formation is strictly terrestric and has a high water-holding capacity and water retention, which means that it is almost never completely dry. With loss of humidity, the humus mass shrinks and forms irregularly shaped shrinkage cracks, as can be seen here. The humus form develops from the alpine Proto-Rendzina (2C) as soon as the vegetation (mainly *Carex firma*) becomes a dense sod. Decomposition and humification are intensified, and at the same time an effective solution weathering takes place, dissolving the clastic calcite débris and leaving only a few, mostly rounded, grains. In this process the formerly abundant plant splinters and calcium carbonate débris disappear and are almost completely replaced by masses of coprogenous constituents.

F.   Droppings of *Glomeris hexasticha,* southern Vienna Woods, Austria. This millipede, similar to a wood louse in form and behavior, takes up considerable amounts of mineral soil with its food, as do almost all the large arthropods. Their droppings or fragments may predominate in a humus form and produce a structure with considerable aggregate formation. The binding materials in these aggregates are humus substances. They are highly stable when wetted, but break apart easily under mechanical stress. Humus forms of this composition on limestone are called mull-like rendzina moder. The droppings seen here are about 2 mm. long. A kind of haft is visible on one end.

Plate 3

TERRAE CALCIS VARIETIES

A.   The (B) horizon of a Terra fusca, southern Vienna Woods, Austria. This soil has a typical braunlehm fabric with a dense matrix traversed by shrinkage cracks; it contains only a few quartz grains because of the advanced state of weathering of the soil mass. A brown concretion of iron hydroxide with a smooth surface is visible at left center. The braunlehm character of this polygenetic formation can generally be found in the highly dispersed soil remnants contained in the limestone from which it is freed by solution weathering. 60×.

B.   The Terra fusca of the southern Vienna Woods of 3A seen under crossed Nicols. Mica-like clay minerals that predominate in the matrix are responsible for the double refractive streaks produced by particle arrangement. An opaque concretion of iron hydroxide can be seen at left center.

C.   An earthy Terra fusca of the Balkan Mountains, Bulgaria. Like other braunlehm fabrics, the fabric of the Terra fusca causes the previously peptized and amorphous iron hydroxide to flocculate in the matrix. This occurs with increasing altitude and a more effective alternation of warm and cold temperature relations. The matrix becomes earthy and deep brown in color,* and, with increasing aggregate formation, it develops a loose structure with high pore space content, good aeration and favorable water conditions. The large fragment of dark-brown iron hydroxide concretion in the lower part of the picture was formed under environmental conditions antedating those of the limestone, that is, at a time when iron hydroxide was peptized and diffused. The limestone fragment above the concretion contains the original peptized iron hydroxide, which was protected against flocculation inside the rock structure.

D.   The preparation of 3C seen under crossed Nicols. Double refractive streaks of the former peptized braunlehm matrix can be observed in the matrix, within the flocculated iron hydroxide. The effect of the particle arrangement is especially noticeable on the surface of the iron concretion (below) and the rock fragment (above). The parallel arrangement of the clay crystals is produced in their deposition. From the low polarization color of the rock fragment, it is evident that it has been transformed into chert as a result of impregnation with colloidal silica which gradually crystallized. The calcium carbonate disappeared completely by solution. This transformation is very typical in the Terra fusca of the humid tropics and also of parts of the subtropics.

E.   A tropical Terra fusca, Lower Congo. The matrix has the fabric of an equatorial Braunlehm, with a clay substance consisting mainly of halloysites and metahalloysites as clay minerals. Under crossed Nicols it generally exhibits no double refractive streaks, as in 3B; this is because

* There is too much red in the color print. The true color of the groundmass is Munsell 2.5 YR 4/4; the yellow of the limestone fragment is 10 YR 8/4.

of the form and small size of the clay mineral crystals and their low double refraction. The matrix contains peptized amorphous iron hydroxide and hydrated colloidal silica as protective colloid. In the cracks of the mineral grains there are numerous precipitations of crystallized iron hydroxides that have been produced by pseudogleyzation or laterization.

F. A Terra rossa with rotlehm fabric, southern Spain. As in any other Rotlehm variety, the red color is produced by the transformation of the amorphous iron hydroxide of the braunlehm fabric into crystallized forms, mainly goethite and hematite. In comparing the micromorphology of Rotlehm varieties (produced by rubification) with that of laterite formations (Plate 12), it is evident that the genesis and fabric character of the two types are fundamentally different. Since the rubified soils are strictly terrestric, having been formed in an alternately wet and humid tropical or subtropical climate, the crystallization is produced in the form of tiny crystallites; these are generally visible only under the electron microscope. The reddening of the groundmass is not produced in the form of stains or concrete crystal aggregates. The crystallites are imbedded singly in the matrix and thus produce a homogeneous coloring of the soil mass in rubified horizons.

Plate 4

HUMUS FORMS OF SOILS ON SILICATE ROCKS

A. Raw humus of a spruce forest, Iron Podzol, F horizon, Hamburg, West Germany. The droppings of horn mites in the feeding cavity of the spruce needle have the same yellow color as the cell walls of the needle. Decomposition and humification of plant residues and animal droppings are very low.

B. Enlarged section of a spruce needle in a raw humus containing red droppings of horn mites. The presence of the same red color of lignin derivates in the undisturbed cell tissue of the needle indicates unfavorable conditions for humus formation.

C. A silicate moder of a Proto-Ranker in beech forest, Styria, Austria. The constituents are little-decomposed plant splinters, loose mineral grains and blackish droppings of small arthropods. Aggregate formation is lacking or imperfect.

D.   Silicate moder of an alpine Sod Podzol, grassland belt, Austrian Alps. This humus form contains reddish plant residues, loose mineral grains and blackish droppings of small arthropods. There is no aggregate formation.

E.   A mull-like moder on granite, Lower Austria. This humus form displays considerable aggregate formation but incomplete decomposition of a part of the plant residues. The main binding material consists of humus substances. Clay is lacking because of incomplete chemical weathering.

F.   Mull of a Chernozem on loess, Ukraine, U.S.S.R. This humus form displays complete disintegration and humification of the plant residues in consequence of a highly developed soil life of great variety, including bacteria and small and large arthropods. The process is finished by the intense activity of mull-forming earthworms. The structure is produced by coalescent droppings of earthworms that form a sponge structure with high water-holding capacity, good aeration, and abundant space for organisms that require considerable oxygen and humidity. Mull formation with sponge structure means that the organisms are able to create their own favorable soil climate, that is, a climate that is more humid than the climate above the soil surface (higher humidity of the soil air, lower dew point). The main binding substances of the aggregate are mixtures of clay minerals, iron hydroxides, colloidal silica and humic substances.

Plate 5

ANMOOR-LIKE FORMATIONS

A.   Eutrophic Anmoor, Bavaria, West Germany. Humification is comparatively good, the structure is dense and humus substances are variously concentrated. Note the presence of blackish coal-like fragments with partly preserved cell structures.

B.   Semiterrestric polar mud soil, Spitsbergen, Norway. Soil aggregates and isolated sand grains are encased in typical mud coatings produced by the precipitation of peptizable substances out of the soil solution; *in statu nascendi* all spaces between the aggregates of this soil are completely filled.

C.   Dystrophic Anmoor, Padul, Andalusia, Spain. The soil fabric is dense, and the organic substances are variously concentrated. In addition to undecomposed plant remnants impregnated by dark-brown and blackish humus substances, there occur reddish to ocherous water-soluble humus substances which accumulate in conducting channels (center). After drying, these substances harden and form shrinkage cracks.

D.   Detail of dystrophic Anmoor of 5C, showing little-decomposed plant remnants impregnated by dark-brown to blackish substances. In the upper half of the picture there are accumulations of ocherous water-soluble humus substances interspersed by shrinkage cracks.

E.   Fossil anmoor-like, tirsoid humus formation on loess (Pseudo-Chernozem), near Stuttgart, West Germany. The soil fabric is very dense and has undergone strong chemical weathering. The clay content is high, and shrinkage cracks are wide after the soil dries out. Typically anmooric humus substances accumulate, some of them blackish in color.

F.   A/B horizon of the Pseudo-Chernozem of 5E. The strong chemical weathering has produced a dense lessivé fabric. In spite of the dense groundmass, the fabric is a braunerde fabric with fillings of braunlehm fabric in the conducting channels (which are seen here in horizontal cross section). These fillings also contain dark humus substances in peptized form in the braunlehm plasma. In the process of peptization they can be transported with the braunlehm plasma deep into the B horizon, along its conducting channels.

Plate 6
BRAUNERDE VARIETIES*

A.   Brown alpine Råmark, Mula Pass, at 5,600 m. (18,370 ft.), Himalayas. The mineral grains are partly encased in a highly earthened coating. Some braunerde plasma fragments appear between the grains.

B.   Alpine Sod Braunerde, (B) horizon, below Mula Pass, at 5,000 m. (16,400 ft.), Himalayas. The braunerde fabric is highly earthened and stained with flocculated amorphous ferric hydroxide. The aggregates (formed by animal activity) are combined into a loose but water-stable sponge fabric.

*The soil samples of 6A and 6B were collected by Dr. Claus Kubiëna, those of 6D and 6E by Dr. S. Kowalinsky.

C. Humid Central European Braunerde on gneiss, (B) horizon, western Hungary. The highly earthened sponge fabric has high water stability and is rich in pore space.

D. Semihumid Braunerde on loess, A horizon, northwest of Peking, China. The sponge fabric with high water and air capacity displays a variety of morphology in its aggregate and space formation. The excellent mull humus is completely decomposed and humified. Organic and mineral plasma constituents are combined with clay humus complexes. At lower left is a feeding cavity in a highly decomposed plant residue filled with small, well-humified droppings of oribatids.

E. Semihumid Braunerde on loess, (B) horizon, northwest of Peking, China. The well-developed sponge fabric has rich aggregate formation and displays throughout a highly earthened groundmass stabilized by flocculated amorphous iron hydroxides.*

F. Mediterranean Dry-Braunerde under *Quercus tozza* on granite, (B) horizon, Sierra de Guadarrama, Spain. Altitude under 1,000 m. (3,300 ft.). The alternation of hardened braunlehm plasma and flocculated braunerde plasma produces a dense soil mass rich in fabric skeleton.

Plate 7

BROWN EARTHENING OF BRAUNLEHM FABRIC†

A. Equatorial Braunlehm on gneiss under virgin forest, (B) horizon, Río Muni, Equatorial Guinea. The dense braunlehm fabric is stained with peptized amorphous iron hydroxide. The only space formations are shrinkage cracks. Strong chemical weathering has left only quartz grains in the fabric skeleton. Dark-brown concretional precipitations of amorphous iron hydroxide are visible in some of the cracks.

B. Earthy Braunlehm-Lessivé from basalt, (B) horizon, Mount Cameroon, equatorial West Africa. The matrix displays brown granulation throughout. This is a young formation in which the process of brown earthening is in its initial stage. The fabric is still very dense and is

---

* The true color is a deeper brown, or Munsell 7.5 YR 5/6.
† The soil samples of 7C, 7D and 7E were collected by Dr. S. Kowalinsky.

subject to great wilting and shrinking, so that the matrix is traversed by a large number of shrinkage cracks. In the center is a large, well-rounded iron hydroxide concretion that was formed in a pre-stage of development when the matrix was fully peptized. Part of this older, highly peptized, mobile, bright-yellow braunlehm plasma can be seen in some of the cracks. These represent old cracks of the fabric and are comparatively wide because of the high shrinkage capacity of the soil mass in the earlier stage. The recent cracks are narrower and completely devoid of braunlehm plasma. A decomposing fragment of plagioclase can be seen in the upper middle part of the picture. On its left are the typical contours of solution weathering, to which silicates are also subject in the ever-humid tropics.

C. Braunerde developed from Braunerde-Lessivé on loess, (B) horizon, northwest of Peking, China. This soil has a braunerde morphology with a sponge fabric rich in pore space. The matrix is well stabilized by flocculated amorphous iron hydroxide. A dark-brown band that crosses the soil mass diagonally traces an older conducting groove now darkened by a brown granulation of precipitated amorphous iron hydroxide. A similar concentration is seen at lower right. A reddish-ocher remnant of the braunlehm plasma is visible in a little crack.

D. The preparation of 7C seen under crossed Nicols. The double refraction shows up the remnants of the older braunlehm plasma in the conducting groove in the center and in the conducting groove fragment at lower right. The details of the granulation are much clearer, as is also the difference between fine quartz sand grains (white) and spaces (black), both of which look white in 7C.

E. Braunerde developed from a Pseudogley-Lessivé on loess, (B) horizon, northwest of Peking, China. The fabric consists of large fragments of the older Pseudogley-Lessivé which have combined into a stable sponge fabric rich in spaces. The following fragmentary constituents of the older fabric can be recognized: (1) traces of iron hydroxide stains, now highly earthened; (2) light-colored remnants of the groundmass of the Pseudogley, now also stabilized by some brown granulation; (3) stratified reddish-brown fragments of space fillings of conducting grooves, which also exhibit dark-brown granulation.

F. Cave loam, from a cave entrance, limestone Alps, Lower Austria. Most of the loam accumulations in the caves of calcareous mountains represent sediments of the present or former soil cover outside the caves. Here the cave loam has the fabric of a transformed Terra fusca. The amorphous iron hydroxide peptized earlier is for the most part precipitated. At upper left are round, dark-brown concretions that have been produced in the way they usually are under warm, humid climatic conditions. Concretionary precipitations, somewhat less separated into rounded individuals, also appear at lower right. In the center, however, there are precipitations whose morphology approaches brown granulation and brown earthening, processes that are typical for temperate and cool climates with cold winters. Between the precipitation types in Braunlehm, on the one end, and Braunerde, on the other, are the transition forms that indicate that the two processes follow the same principle. The differences lie in the fact that large round concretions (which may attain the size of a child's head in tropical Braunlehm) can be formed only with constant temperature and humidity, without the interruptions of repeated frost or dryness. Here only very small though very numerous precipitation bodies can be formed, that is, accumulations of numerous concretion nuclei instead of a few well-developed single concretions.

Plate 8

EARTHY BRAUNLEHM

A.   Earthy Braunlehm on basalt, (B) horizon, Fernando Póo, Equatorial Guinea. Altitude 600 m. (1,970 ft.). The matrix, instead of being a dense fabric with great wilting and shrinking capacity, is completely subdivided into irregularly formed aggregates. The dense matrices of the aggregates are partly preserved, and in them amorphous iron hydroxide is precipitated, the dark-brown color giving way with increasing earthening of the whole soil mass.*

B.   Earthy Braunlehm on Braunlehm sediment, (B) horizon, Sierra Leone, West Africa. The dense soil mass is subdivided into aggregates that contain flocculated iron hydroxide in their interior. Part of the older yellow braunlehm matrix is visible between the precipitates. Some of the aggregates have blocky forms and sharp delimitations, indicating less water stability.

C.   Vega of earthy Braunlehm, A horizon, banana plantation region, Río Frío, District of Santa Marta, Colombia. This soil has a very stable

---

* The true color is Munsell 2.5 YR 3/4 (less red).

sponge fabric with high water and air capacity. The grains of the fabric skeleton are connected by thin bridges of earthy fabric plasma. This soil is rich in well-preserved silicates, such as orthoclases, plagioclases, biotites, hornblendes and augites. The excellent mull formation, together with the characters above mentioned, explains its high fertility.

D.   Fabric of 8C seen under crossed Nicols. The distinction between colorless mineral grains and the spaces is much clearer than under ordinary light. Also the nature of the sponge fabric with its thin fabric bridges shows up well.

E.   Earthy Braunlehm on gneiss, A horizon, Greenville District, Liberia, Gulf of Guinea. The brown earthening of this soil was induced by human activity, primarily by mulching and the application of compost. Both procedures promoted the development of soil life, especially the activity of the soil animals. The aggregate forms are still blocky, but their interiors contain a dense, dark-brown granulation of amorphous iron hydroxide.

F.   Earthy Polvillo-Braunlehm on basalt, A/(B) horizon, near Moka, Fernando Póo, Equatorial Guinea. These typical Polvillo aggregates are formed in soils of high plasticity whose aggregates are highly water stable. They are produced in horizons densely populated by soil animals. Their forms become rounded as a result of the continuous rolling movements of the numerous soil inhabitants.*

Plate 9

RUBIFICATION AND XEROMORPHIC PULVERIZATION OF BRAUNLEHM FABRICS

A.   Rotlehm, (B) horizon, Georgia, U.S.A. The fabric is dense, and shrinkage cracks are the only kind of spaces in the matrix. It differs from a braunlehm fabric only in that the red color is homogeneous, being produced by the precipitation of very fine crystallites of goethite or hematite in the fabric plasma (rubification). As a result of strong chemical weathering, the fabric skeleton consists almost entirely of quartz grains.

B.   Earthy Rotlehm, (B) horizon, Kenya, East Africa. The goethite and hematite crystallites have flocculated in the yellow braunlehm matrix (red earthening). The transformation into the earthy variety is in its beginning. The still rather dense fabric is interspersed by numerous

* The true color is Munsell 2.5 YR 3/4 (less red).

shrinkage cracks, but the cracks are very irregular in form, width and direction, as a consequence of irregularities in the matrix produced by different degrees of earthening.

C.   Earthy Rotlehm, (B) horizon, Kenya, East Africa. An entire thin section is shown at low magnification. As a result of the earthening, only a few spaces in the form of shrinkage cracks are left. The majority of the spaces are rounded forms with irregular delimitation (especially at upper right).

D.   Partly pulverized Braunlehm sediment relict in the dry desert, Tunisian Sahara. This relict displays a typical transformation of braun-lehm fabric on the soil surface or on the surface of cracks as a result of alternate wetting and complete drying in conjunction with extreme heat-ing. The sediment was deposited under humid conditions. The strong weathering has left only quartz grains in the fabric skeleton. The inter-granular spaces are filled with braunlehm plasma, which forms an almost continuous plasma layer on the older surface (compare 9E). Above this layer is the pulverization zone, consisting entirely of loose constituents, that is, of quartz grains (generally covered with a fine braunlehm coat-ing) and angular blocky fragments of the former braunlehm matrix and fine dust of braunlehm plasma (top center).

E.   A larger field of the pulverized Braunlehm sediment of 9D seen at lower magnification. The three characteristic layers are more evident. The unaltered Braunlehm sediment (lower half)* is followed by the plasma layer formed as a coating on the older soil surface. On top is the pulverized layer, with elements partially disengaged from other surface parts and deposited over the plasma layer.

F.   Another view of the soil of 9D seen under crossed Nicols. The nature of the fabric skeleton and the fabric plasma is very clear. The latter, although disturbed by erosion, still has the double refraction of the plasma fragments and the coatings of the grain surfaces.

Plate 10
LESSIVÉ VARIETIES

A.   Braunlehm-Lessivé on loess, B horizon, North Island, New Zealand. Every lessivé fabric is two-phased, in that the plasma types in the matrix

---

* The true color is Munsell 2.5 YR 4/4 (less red).

and in the conducting grooves are different. The plasma types in this soil have the same micromorphology; both have a typical braunlehm fabric (hence the name Braunlehm-Lessivé). The difference between the plasma of the matrix and that of the fillings of the conducting grooves is that the latter has much more mobility and therefore is more easily washed out and moved into the empty conducting grooves (shrinkage cracks, root channels and earthworm tubes). The plasma of the matrix has lost mobility because the environmental conditions of the habitat changed from a humid to a semidry climate. The fillings of the conducting grooves do not contain fabric skeleton.

B.   Rotlehm-Lessivé on gneiss, B horizon, La Vera, central Spain. This soil has a dense, slightly earthy rotlehm fabric in the matrix, but yellowish, very mobile and highly stratified braunlehm plasma in the fillings of the conducting grooves. Braunlehm concretions that were formed prior to the rubification of the matrix appear at right.

C.   Red Pseudogley-Lessivé on biotite gneiss, g/B horizon, Sierra de Gredos, central Spain. The matrix is considerably de-ironized, containing almost undecomposed crystals of augite and slightly browned biotite. There is bright-red fluidal plasma of Pseudogley origin in the conducting grooves (compare Plate 15E and 15F).

Plate 11
PSEUDOGLEYZATION

A.   Stagnogley under spruce forest on loess, G horizon, Weser Bergland, West Germany. This true Gley, produced by stagnating surface water, has a dense fabric in which all ferric hydroxide has been reduced. Gradually a secondary precipitation of ferric hydroxide takes place by oxidation of the ferrous compounds, producing scattered dark-brown stains with irregular delimitation.

B.   Sediment relict of Braunlehm with beginning pseudogleyzation on loess, Invercargill, South Island, New Zealand. Without a reduction, a red stain is produced by crystallization of amorphous ferric hydroxide. The stain has been deformed by slope movement of the soil.

C.   Braunlehm on graywacke with beginning pseudogleyzation, (B) horizon, Taita, North Island, New Zealand. The pseudogleyzation starts with the formation of concretion-like stains of goethite. These are produced by the crystallization of peptized amorphous ferric hydroxide that migrates by diffusion to the precipitation centers.

D.   Pseudogley from Braunlehm on loess, g horizon, Otautau, South Island, New Zealand. Here the concretion-like goethite stain displays a further stage of development. The former braunlehm matrix has completely bleached as the amorphous ferric hydroxide moved away to concentrate in the crystallization zone. The red stain has lost its former sharp delimitation and is developing towards new forms.

E.   Pseudogley in Braunlehm sediment, E/g horizon, Greytown-Martinborough Road, North Island, New Zealand. The precipitation of goethite from amorphous ferric hydroxide is in a progressive stage.

F.   Pseudogley in Braunlehm sediment. E/g horizon, Invercargill, South Island, New Zealand. The sediment has a dense, light-yellow (because of some loss of iron) braunlehm fabric with irregular cracks and rounded spaces. The crystallization in this fabric is not produced in the form of stains but rather in the form of marginal concretionary precipitations of goethite on the crack or space walls. The formation of goethite is established, not by the oxidation of ferrous compounds dissolved from the interior of the spaces, but by crystallization of peptized amorphous ferric hydroxide that migrates from the interior of the matrix to its margins, where it finds space to crystallize more easily. All these precipitations tend to grow from the space walls into the interior of the matrix rather than into the interior of the cracks or rounded cavities.

Plate 12

LATERIZATION

A.   Lateritic mottled clay, on gneiss, Río Muni, Equatorial Guinea. In a matrix of braunlehm fabric, peptized amorphous ferric hydroxide moves by diffusion towards the precipitation centers; here it crystallizes mainly as goethite or hematite. The precipitation takes the form of stains or marginal coatings (left). With greater density, they become opaque; they appear blackish under transmitted light, but bright red when investigated under vertical or incident light.

B. This dense laterite fabric displays crystallized ferric hydroxide in the form of thin red needles typical of goethite.

C. Cuirass-Laterite from Braunlehm on gneiss, Río Muni, Equatorial Guinea. The old braunlehm matrix contains numerous hardened crystallization aggregates of hematite in a great variety of forms, such as droplets, ledges and stains. The yellow matrix between the ledges and stains is partly preserved, partly washed out (white parts with irregular delimitation). The very dense hematite is opaque and therefore appears blackish under transmitted light.

D. A detail of the Laterite of 12C seen under incident light. The bright-red crystallization aggregates of hematite are very evident.

E. This dense laterite fabric contains ferric hydroxide in the form of goethite, this time in spheroidal crystallization aggregates.

F. Braunlehm-Laterite, Greenville District, Liberia, Gulf of Guinea. Crystallized aluminum hydroxide in the form of gibbsite is visible between ledges of hematite. The braunlehm plasma, in which the gibbsite crystals were formed, has been washed out.

Plate 13
VULCANOGENIC EDAPHOIDS

A. Braunlehm-like edaphoid, Valdivia, Chile. This edaphoid has a perfect braunlehm-like fabric with a very dense matrix that is almost completely free of fabric skeleton. It has a high shrinkage capacity. Note the wide, very regular and straight shrinkage cracks.

B. Trumao clay loam, Valdivia, Chile. This soil illustrates the transformation of a braunlehm-like edaphoid by soil formation in a temperate climate with cold winters. Brown earthening has taken place.

C. Rotlehm-like edaphoid, Vogelsberg, West Germany. This edaphoid has a dense matrix with shrinkage cracks. It is stained by minute hematite crystals such as are produced in tropical soils by rubification.

D. Pseudogley-like edaphoid, Valdivia, Chile. Blocky aggregates with sharp delimitation are contained in a braunlehm-like fabric with dazzling red hematitic stains.

E. The rotlehm-like edaphoid of 13C seen under crossed Nicols. Some double refractive streaks produced by particle arrangement are visible in the fabric.

F. Laterite-like edaphoid, Vesuvius, Italy. The fabric of this edaphoid resembles that of lateritic mottled clays. It has a whitish matrix with red and yellow stains of crystallized iron hydroxides.

G. Rotlehm-like edaphoid, Vogelsberg, West Germany. The dense, bright-red matrix contains a large number of circular concretions such as are produced in a braunlehm fabric (dark-brown amorphous iron hydroxide). During the development of rubified soils they become bright red as a result of transformation into hematite. At upper right a large, well-rounded concretion with sharp delimitation is visible. Because of the great density of the concretions in the edaphoids, they are opaque and appear blackish when investigated under transmitted light.

H. Laterite-like edaphoid, Vogelsberg, West Germany. The fabric of this edaphoid is not only very similar to the fabric of true Laterite but is capable of hardening and forming volcanic ironstones.

Plate 14

LOESS SOILS

A. Fossil Braunlehm sediment, Otautau, South Island, New Zealand. This sediment displays a dense soil mass with highly peptized fabric plasma and wide shrinkage cracks. Here the development of a lessivé fabric is in its beginning (compare the accumulation of fabric plasma in a soil space in the lower part of the picture). A fragment of the matrix of old Braunlehm sediment is visible at upper left (compare Plate 15A and 15B). At right is an iron hydroxide concretion that represents a typical phenomenon of the braunlehm dynamics. 60×.

B. Braunerde containing irreversibly hardened flakes of braunlehm plasma, (B) horizon, altitude about 1,000 m. (3,300 ft.), South Island, New Zealand. The presence of irreversibly hardened braunlehm flakes is a common phenomenon in some loess areas (see 14A, above, and Plate 15A, 15B, 15C and 15D). The stability of the flakes in this fabric is remarkable because the soil is a humid Braunerde and almost no traces of brown granulation are found in the interior of the flakes. The stabilization of this flake took place before the formation of the Braunerde.

C.   Braunerde-Lessivé, A_e horizon, Laupheim, West Germany. Here is the typical fabric of the eluvial horizon of a Braunerde-Lessivé. The well-rounded, dark-brown concretions of iron hydroxide indicate that the former fabric was chiefly a braunlehm fabric. After some brown earthening all the Braunlehm component was washed into the B horizon and filled the conducting grooves. There remain only the Braunerde components, the concretions and the fabric skeleton.

D.   Braunerde-Lessivé, B horizon, Laupheim, West Germany. The matrix contains braunerde plasma, and the conducting grooves are completely filled with a highly peptized brownish-ocher braunlehm plasma that is almost completely free of fabric skeleton. It belongs to the oceanic section of the humid temperate subzone in Western Europe, where both high precipitation and mild winters favor the preservation of the braunlehm fabric and the formation of fluidal plasma. The widest shrinkage cracks, root channels and earthworm tubes become completely filled up, without any microstratification in the braunlehm plasma. 60×.

E.   Filling of braunlehm plasma in a wide conducting groove (shrinkage crack) of the B horizon of Braunerde-Lessivé, near Heilbronn, West Germany. The crack was not only completely filled, but the braunlehm plasma had not developed even the slightest traces of brown granulation; a large number of fine cracks was observed. 60×.

F.   Relict of braunerde-lessivé fabric in Chernozem, A/B horizon, Kharkov, Ukraine, U.S.S.R. The fabric displays mull formation, but highly developed brown earthening is apparent in the matrix. Both are conducive to the formation of a sponge fabric. Some of the spaces contain very small stratified remnants, as well as greatly shrunk and inactive fillings of braunlehm plasma. The latter are hardened and display no brown granulation. These fillings, in contrast with the braunlehm fillings in 14D and 14E, are typical for a number of loess soils in the semiarid steppe zone.

Plate 15

RAW LOESS VARIETIES

A.   Tertiary marine marl with fillings of Braunlehm sediment in its cavities, Invercargill, South Island, New Zealand. Fragments of these fillings in an aged and irreversibly hardened state are common in the loesses and loess soils of the island. It can be assumed that a large part of the reactivated braunlehm plasma in these soils was derived from braunlehm-rich sediments that dried out and became subject to wind erosion during geological periods when the sea regressed and there were dry interglacials.

B. Hardened and inactivated fragment of braunlehm fabric in a Braunerde on loess, South Island, New Zealand. This fragment looks very much like the braunlehm fillings in the cavities of the marine marl in 15A.

C. Raw loess, West Germany, seen under incident light. This kind of light shows up constituents that are barely visible under ordinary transmitted light. Braunlehm sediments previously eroded and transported by wind action are present in this preparation in the form of yellow or orange angular to rounded fragments of a peptized braunlehm matrix or a pseudogleyed braunlehm plasma; or they exist in the form of braunlehm coatings or braunlehm deposits on the surface of the mineral grains.

D. Raw loess, West Germany, seen under incident light. Angular fragments of braunlehm plasma and mineral grains, with braunlehm coatings and deposits on their surfaces, can be seen in this preparation. With increasing humidity another phenomenon takes place: percolating water causes the braunlehm flakes to lose their delimitations by flowing apart and forming irregular excrescences.

E. Raw loess, South Island, New Zealand. Reactivated braunlehm plasma has begun to move and to be deposited in the form of coatings of fluidal plasma on the walls of the conducting grooves of the raw loess. The partly pseudogleyed material attained a uniform color after complete mixing. Thus a kind of lessivé fabric can be produced in a raw loess in a habitat that offers neither conditions for pseudogleyzation nor indications for advanced weathering.

F. Raw loess, South Island, New Zealand. The preparation of 15E seen under crossed Nicols. A striking double refraction of the fluidal plasma caused by particle arrangement can be observed, and this is also evident in some of the intergranular spaces. Thus this raw loess has undergone a development that is possible only with highly weathered soils in humid tropical or subtropical climates (assuming monogenetic soil development).

Glossary

aggregate (Dumont, 1913): Small, loose structural bodies of various origin, formed in nature; often called soil crumbs. Aggregates may be either true crumbs (with a flaky fabric and an irregular scabby surface), or small fragments, or small clods, or droppings of small animals.

allochthonous: Of or relating to soils, soil constituents or soil characteristics that developed in another location in an earlier period.

bole, bolus: Any of several varieties of very dense clay, rich in silica and displaying a conchoidal fracture; dazzling ocher-yellow, brown, red, gray, bluish or white in color and very waxy to greasy-lustrous in appearance. Boluses are derived primarily from vesicles in basalts, amygdaloids and cavities in dense limestone. They have medical and technical uses.

classification: An ordering of the most accessible objects in the realm of soils (for example, the arable soils of a country), beginning with the specialized and gradually progressing to the general; objects are sorted into classes already established or to be created. Generally, classification of soils is designed to serve branches of agriculture, forestry, engineering and other special fields, or is limited to specific administrative regions. It is the reverse of *division* (*systematics*).

climax soil: The maximum attainable end phase of a soil development sequence in a given location. The kind of climax soil is largely dependent on the environmental conditions of the habitat. In dry deserts, the desert raw soil is both the initial and the final phase of the sequence. In the Podzol zone on calcareous sandstone, the climax soil (Iron Podzol) may be reached in a development sequence of eight phases (carbonate raw soil, Proto-Rendzina, mull-like Rendzina, Mull-Rendzina, brown Rendzina, Braunerde, podzolic Braunerde, Iron Podzol). In landscapes subject to strong soil erosion, the climax formations of many-phased development sequences usually do not occur alone but rather in conjunction with earlier phases of the development sequences.

conducting channels: The channels in a dense soil fabric that allow downflow of percolating water. Conducting channels are usually somewhat vertical, and most represent drought fissures, earthworm tubes and root channels. Substances dissolved or washed down in percolating water are deposited in a characteristic way in conducting channels. Hence the content of the channels has importance diagnostically. The most common fillings are calcium carbonate, humus substances and highly peptized inorganic colloids.

237

development sequence: Synonymous with genetic soils series (Pallmann) or development series. In a soil development sequence (soil types, subtypes or varieties), each soil is derived from its predecessor, and all soils in the sequence progress from the primitive to the complex (for example, limestone raw soil, Proto-Rendzina, mull-like Rendzina, brown Rendzina, Terra fusca, Terra rossa). Many development sequences suffer a loss of fertility in the end phases. This is known as aging.

diagnosis: The recognition and naming of soil formations on the basis of characteristic and distinctive properties, beginning with characteristics that are easy to determine.

division (systematics): The systematic arrangement of a totality (kingdom) of soil formations by the creation of graduated concepts, starting with the most general categories and progressing through intermediate categories to the most specialized. Division is the reverse of *classification.* The latter does not start from the whole and does not necessarily consider the whole, but creates special units (classes) and advances gradually to the most general.

droppings: That part of the humus which is composed of the excreta of small animals. Droppings may be well-formed or constitute residues. They are usually dark in color and display stronger decomposition and humification, as well as more complete mixing and combination of organic and mineral substances, than do other parts of the humus.

dystrophic: Of or relating to humus and soil formations that have very unfavorable biological relations and are tolerated only by some organisms.

eluviation: The removal of soil substances by percolating waters that move downward in the soil.

erodibility: The intensity of soil removal by wind, running water or moisture that causes sliding, given the same conditions (the same amount of rainfall, slope inclination, wind force, protective vegetation, etc.).

eutrophic: Of or relating to soils with high nutrient content and high biological activity.

fabric: The structural arrangement of the soil constituents. Used not only in the narrower sense to refer to aggregate formation but also in a general sense to refer to the inner fabric of dense soil masses, the effect of the processes of precipitation and solution, the movement of substances, the alterations caused by organisms, and the like.

fossil soil: Soil that is fossilized, dead, buried, petrified or diagenetically altered (hardened by cements, transformed into coal or carboniferous sandstone by carbonization). As opposed to *recent soil* and *relict soil.*

fulvic acids (Odén, 1912): A group of very different acid-soluble humus substances that can be extracted from the soil by leaching with dilute alkalies, but, unlike *humic acids,* cannot be precipitated with mineral acids. The fulvic acids constitute the major part of the acid humus sols of the Podzols and Semipodzols. Although the concept of fulvic acids embraces a group of humus substances that are not homogeneous and their composition is little known, the concept is an essential aid in soil diagnosis.

grain separation: A peculiarity of highly erodible, easily silted up soils. It consists in complete destruction of the original soil fabric by erosion and subsequent sedimentation and separation of the various textural fractions. The fractions are sorted out during transportation according to the velocity of the running water. Thus there may form a series of layers of different colors (Plate 14), beginning with the coarse sand fractions and ending with extremely fine colloidal substances. Grain separation occurs in soils susceptible to erosion even during moderately heavy rains and even where there are only slight differences in relief in ploughed fields.

humic acids (Odén, 1912): A group of blackish-brown nitrogen-containing humus substances that can be extracted from *humus* by leaching with dilute alkalies and then can be precipitated with mineral acids. With calcium ions they form blackish compounds (calc humates) that are difficult to dissolve. Such compounds are characteristic of the humus of the Rendzinas. The proportion of humic acids is very high in true mull formations. Here they (together with iron hydroxide and silica) appear as dyes in the clay substance which do not separate easily (hence, the occurrence of clay-humus complexes and other mineral humus complexes).

hydrophilic colloids: Colloidal substances that are little subject to flocculation by electrolytes in dispersions. They have a strong affinity for water and on thickening form a jelly-like substance. They are found as humic substances, tannic substances and hydrated colloidal silica in the soil.

hydrophobic colloids: Colloid substances that are easily flocculated by electrolytes. They lack an affinity for water and do not form a jelly on thickening. They are found as ferric hydroxide and aluminum hydroxide in the soil.

illuvial horizon (Wyssotski, 1901): An enriched horizon or soil layer containing substances that were deposited in it or washed into it as they were removed from other parts of the profile by solution or dispersion. Either percolating water or mobile capillary water can produce the enrichment. With strong capillarity, illuvial horizons can be formed in the surface layer of the soil. The most frequent types of illuvial horizons are lime-enriched horizons (Ca horizons), salt-enriched horizons (Sa horizons, including gypsum-enriched or Y horizons), raseneisenstein or raseneisenerde layers in gley soils (Fe horizons), humus orstein or humus orterde layers (B_h horizons), and sesquioxide-enriched layers (B_s horizons) in Podzols.

microskeleton (fabric skeleton): The coarse, little-weathered mineral particles or little-decomposed and little-humified organic constituents of a soil fabric. Visible in thin section preparations or with direct microscopy.

natural system: In *division* (*systematics*), the ordering of objects according to all their characteristics, but not on the basis of just one or a few properties (as in the so-called artificial system). The arrangement of the units is governed, not by an arbitrary principle of division, but by their mutual connection and inner relationship, that is, according to their essential order in nature.

oligotrophic: Of or relating to soils with low nutrient content and relatively low biological activity. Oligotrophic soils are usually formed on base-deficient parent rocks.

organic substance: The totality of the organic constituents of the soil, including all undecomposed organic remains and all macroscopically inseparable living organisms. Whereas the content of organic substances and, with certain restrictions, the content of *humic substances* (true humic substances) are directly determinable chemically in a given soil, the content of the constituents of the humus can only be investigated microscopically at the present time. This limitation is unfortunate in view of the importance of the humus constituents in identifying humus forms.

peptization: The more or less complete dispersion (sol formation) of colloidal particles in liquids (particles of 1 to 100 mμ in diameter). The opposite of flocculation (pectization, coagulation), which means the precipitation of particles in flaky aggregates or coagulates.

phenology: A branch of science dealing with the relations between climate and periodic biological phenomenon. Specifically, the characteristic seasonal changes that take place in the soil as a whole, not only with respect to climate (see *soil climate*) but also with respect to physical and chemical properties (*p*H value, salt content, and so forth), structure, and the composition and activity of soil flora and fauna.

recent soil: A soil formed under present environmental conditions.

relict soil: A soil that still displays many essential characteristics of its origin as a soil that developed under the climate and the environmental conditions of an earlier geological period. It forms the living superficial layer of the earth rind in its present location. Relict soils are usually characterized by high stability, for example, Roterde and Rotlehm. Or they stand higher in their development phase than the *climax soil* of the present era; examples are Terra rossa in a region with Xero-Rendzina as the climax soil; brown soils in the region of the Tundra Rankers; and bolus-like silicate soils on old land surfaces in the peak region of the limestone Alps. Not to be confused with *fossil soil.*

rubification: The process resulting in peptization of amorphous iron hydroxide and the formation of tiny crystals of goethite and hematite. The crystals are freely suspended in the dense groundmass of a braunlehm and give it a bright-red color. Rubification is not used in this book in its general sense, that is, to refer to any other process that produces reddening of the soil.

separation (Absonderung): The formation of fragmentary, sharp-edged, angular, platy to prismatic structural bodies in soils with great swelling capacity and in sediments rich in colloidal substances (or in solidified lavas). Separation is caused by the natural breakdown of the entire groundmass as a result of shrinkage.

soil: The transformation layer of the solid earth rind. A soil is inhabited by living organisms and was produced by the environmental conditions peculiar to its habitat. Environmental conditions include such factors as climate, relief, water conditions, plant cover, animal and human activity. A soil is subject to distinctive seasonal changes (*phenology*) and a characteristic development (*soil development*).

soil biology: A branch of soil science that deals with the living organisms and the totality of the vital processes in soils. To be distinguished from *soil dynamics,* which deals with the totality of the nonbiological processes in soils.

soil climate: The average course of changes in the interior of the soil over a period of years, including temperature alterations, variations in the composition of the soil air (moisture and oxygen content) and changes in water relations. The soil climate is subject to great differences by comparison with the air climate (for example, the soil climate of autochthonous Terra rossa, Proto-Rendzina, and low moor); moreover, the same air climate produces great differences within the various soil types or subtypes. Each soil therefore has a characteristic soil climate and this climate has a primary influence on the kind of development it represents. Soil climate is not to be confused with the climate of the air layer of a habitat near the soil, for example, the climate of the vegetation layer, the turf, the herb or the dwarf shrub cover.

soil development: The soil-forming process as it is governed by rules and proceeds in a typical sequence of forms, from the most simple to the most complex and most highly organized, for example, from raw soil, Ranker, Braunerde and podzolic Braunerde to Iron Humus Podzol.

soil dynamics: The totality of the active forces in the soil, including movements and physical and chemical alterations of nonbiological character produced by them (in contrast to *soil biology*).

soil formation: The soil-forming process in general, but not necessarily *soil development* according to rules governing alterations of form or changes of type.

spongy fabric: Biologically favorable soil fabric consisting of aggregates bound to each other in such a way that a system of connected cavities is formed, as in a sponge. The inner structure of the aggregates is generally porous, not dense. Spongy fabric permits good aeration and optimum water economy, creates living space for the non-digging small soil animals, promotes thorough rooting of plants and offers protection against soil erosion.

streaks, double refractive: Very striking irregular, striped to flame-like fabric parts that display double refraction. They are produced by the arrangement of particles in the clay substance, either by the movement of fluids (fluidal structure) or by the deposition of colloidal masses as they dry out in cavities, shrinkage cracks and earthworm channels, or on the surface of granules, concretions and the like (deposition structure).

structure: The manner of formation and arrangement of the structural complexes of the soil (crumbs, gravels, small fragments or other aggregates; also the kind of separation). Structure is used synonymously with *fabric* in the literature of some countries.

Index of Soils

Index of Place Names

About the Author

Walter L. Kubiëna introduced the subject of soil micromorphology to the world of science. He has studied soils on every continent and in over a hundred countries.

On receiving his doctorate in 1927 he joined the faculty of the Department of Geology and Soil Science at the Hochschule für Bodenkultur in Vienna. During the early 1930's he worked with Selman A. Waksman at Rutgers University and at the Centre National de Recherches Agronomiques in Versailles.

His first book, *Micropedology,* describing the method he had developed for the microscopic investigation of soils in thin section, was completed in 1937. Two other major books followed in 1948 and 1953, one summarizing his studies on soil development and the other providing an exhaustive taxonomical treatment of the soils of Europe.

Professor Kubiëna has taught for many years at the University of Hamburg and the Bundesforschungsanstalt für Forst- und Holzwirtschaft in Hamburg-Reinbek. At the invitation of Spanish and Portuguese scientific bodies, he has spent extended periods of time working and lecturing in Spain, Africa, and the East Atlantic Islands. He has been a guest lecturer at a number of American universities and is the recipient of several honorary degrees and numerous awards, including the distinguished Justus von Liebig prize.

The text of this book was set in Caledonia Linotype and printed by offset on Warren's Patina II manufactured by S. D. Warren Company, Boston, Mass. Composed, printed and bound by Quinn & Boden Company, Inc., Rahway, N.J. Color plates printed by Civic Printing and Lithographic Corp., New York, N.Y.